William Dorrance Beach

Manual of Military Field Engineering

For the Use of Officers and Troops of the Line

William Dorrance Beach

Manual of Military Field Engineering
For the Use of Officers and Troops of the Line

ISBN/EAN: 9783337267520

Printed in Europe, USA, Canada, Australia, Japan

Cover: Foto ©Lupo / pixelio.de

More available books at **www.hansebooks.com**

MANUAL

OF

MILITARY FIELD ENGINEERING

FOR THE USE OF

OFFICERS AND TROOPS OF THE LINE.

PREPARED AT THE

UNITED STATES INFANTRY AND CAVALRY SCHOOL

BY THE

DEPARTMENT OF ENGINEERING,

CAPT. WM. D. BEACH, 3RD CAVALRY, INSTRUCTOR.

FORT LEAVENWORTH, KANSAS,
JULY, 1894.

PREFACE.

The necessity existing at the Infantry and Cavalry School for a text book on Field Engineering, including the various military expedients recognized in our service, is deemed sufficient reason for the following pages.

Most of the subjects treated of in this volume may be found in various military works published in our country during the past twenty-five years, but the fact remains that no one book has covered the required ground, nor has their revision been of very recent date; while, at the same time, the new field gun and small calibre rifle have necessarily modified previously existing profiles of Field Works and Shelter Trenches.

Access has been had to corresponding publications of the Germans, French, English and Austrians, as well as to our own Official Rebellion Records and many other available sources, native and foreign.

It has been the endeavor to limit the scope of this work to subjects considered indispensable as a part of a line officer's education.

The following Assistant Instructors in the Department of Engineering, viz:—1st Lieut. E. A. ROOT, 19th Infantry; 1st Lieut. W. C. WREN, 17th Infantry and 1st Lieut. T. H. SLAVENS, 6th Cavalry have been associated with the undersigned in the preparation of this volume.

            WM. D. BEACH,
              Captain, 3rd Cavalry.

U. S. Infantry and Cavalry School,
  Fort Leavenworth, Kansas, July, 1894.

List of Books Consulted in the Preparation of this Work

Aide Memoire, R. E. . . . . *Vols. 1-2.*
A Move for Better Roads . . . . *L. A. Haupt.*
Appleton's Cyclopædia of Applied Mechanics *Vols. 1-2.*
Civil Engineering. . . . . *Wheeler.*
Cours de Fortification Passagere. . . . . *De Guise.*
Ecole de Fortification de Campagne. . . . . *French.*
Elements of Field Fortification. . . . . *Wheeler.*
Engineering News . . . . *Magazine.*
Engineer's Pocket-Book. . . . . *Trautwine.*
Field Fortification. . . . . *Turner.*
Field Fortification . . . . *Hutchinson.*
Field Works. . . . . *Brackenbury.*
Field Works used in War. (Translation
  from the German). . . . . *Wilson.*
Good Roads. . . . . *Magazine.*
Gunpowder and high Explosives . . . . *Walke.*
International Cyclopædia. . . . . *Dodd, Mead & Co.*
Journal of the Military Service Institution
  of the U. S.
Journal of the U. S. Cavalry Association.
Manual for Engineer Troops. . . . . *Duane.*
Manual of Military Engineering . . . . *Ernst.*
Manual for Railway Engineers. . . . . *G. L. Vose.*
Manuel de Fortification de Campagne. . . . *Brialmont.*
Manuel des Travaux de Fortification de
  Campagne par un Capitaine d' Infanterie.
Manuel de Fortification. . . . . *Plessix and Legrand.*
Manual of Heavy Artillery. . . . . *Tidball.*
Military Bridges. . . . . *Haupt.*
Military Bridges. . . . . *Chester.*
Military Engineering. Instruction in. . . . . *Chatham Course.*
Military Transport. . . . . *Furse.*
Modern High Explosives. . . . . *Eissler.*
Notes on Military Hygiene . . . . *A. A. Woodhull.*
Official Records of the Rebellion. U. S. . . . *War Department.*
Organization and Tactics. . . . . *Wagner.*
Pionier Taschenbuch, Berlin, 1893. . . . . *Official.*
Report of Chief Signal Officer. U. S. 1893. *War Department.*
Roads and Railroads. . . . . *Chester.*
Roads and Railroads. . . . . *Gillespie.*
Roads, Streets, and Pavements. . . . . *Gillmore.*
Temporary Fortification. . . . . *Chester.*
The American Railway. . . . . *Scribner.*
U. S. Bridge Equipage and Drill. . . . . *War Department.*

TABLE OF CONTENTS.

Chapter.		Page.	Plate.
I	General Principles............	7	
II	Fire, Projectiles, and Penetration......................	11	
III	Field Geometry..............	14	1, 2
IV	Hasty Intrenchments, Gun Pits and Epaulements...........	23	3, 4
V	Clearing the Ground.........	31	5, 6
VI	Obstacles...................	37	7, 8
VII	Field Works.................	45	9-15
VIII	Working Parties.............	71	16
IX	Revetting Materials and Revetments.....................	74	17, 18
X	Field Casemates and Magazines	89	19, 20
XI	Field Works in Combination...	95	21
XII	Siege Works.................	103	22, 23
XIII	Defense of Localities.........	109	24-29
XIV	Use of Cordage and Spars.....	131	30-33
XV	Spar Bridges................	149	34-40a
XVI	Floating Bridges.............	176	41-49
XVII	Roads......................	212	50
XVIII	Railroads...................	221	51-53
XIX	Telegraph and Telephone Lines	235	54
XX	Demolitions	240	55-57
XXI	Camping Expedients.........	257	58-60

MANUAL

OF

MILITARY FIELD ENGINEERING.

CHAPTER I. General Principles.

1.—**Military Field Engineering** may be defined to be the art of utilizing the materials at hand for the attainment of the security, effectiveness, health and comfort of an army in the field.

The modern rifle has vastly increased the value of cover both in attack and defense, and rendered necessary the application of the principles of fortification to an army in the field. The result to be obtained in all fortification is to so strengthen a position, by artificial means, that a force may successfully resist or subdue another.

2.—**Fortification** is divided into two general classes, viz.;
(a)—**Permanent**.
(b)—**Temporary** or **Field Fortifications**.
With the former this manual has nothing to do.

3.—The latter division includes three quite distinct classes.

The first comprises all works devised for the temporary protection of important points such as cities, arsenals, bridges, fords, positions, etc., and are technically known as *Field Works*.

GENERAL PRINCIPLES.

The second comprises the various devices of the engineer for reducing a fortified place by means of parallels and approaches, called *Siege Works*.

The third division relates particularly to the quickly made defenses by which an army in the presence of an enemy protects itself; these are known as *Battle Intrenchments* or *Hasty Intrenchments*.

4.—A Defensive Position is one affording protection from the shot and observation of an enemy and, at the same time, commanding the ground in front, within range.

A position of perfect defense is not possible, but the following general principles are to be fulfilled as nearly as circumstances will permit.

(1) The defenders' position should conform to the special tactical requirements of the occasion* and should be such as to favor the use of their relatively strongest arm.

(2) It should be made impossible for the enemy to obtain natural cover during his advance. In other words, the position should have a free field of fire.

(3) The defenders should be protected from the fire and view of the enemy by cover so arranged as not to interfere with counter attacks.

(4) The advance of the enemy should be hindered by obstacles so arranged that he may be checked while under the fire of the defenders.

(5) Communications should be such that the defenders may freely move from one part of the position to another, while the contrary should obtain with respect to the enemy's ground in front.

The chief requisite of a defensive position is a free field of fire, especially at short and mid ranges. If the position is judiciously selected the field of fire will generally be obtained without much difficulty, but the advantages of the position and the effect of the fire may be enhanced by temporary fortifications. The cutting down of slight ridges which might afford cover for the enemy within effective range or the removal of hedges, fences, etc., may

---

*A purely defensive position, for instance, might have its flanks resting on impassable obstacles, and thus be secure from a turning movement, but this same position might be found to be a faulty one were a quick offensive movement, by the defenders, contemplated.

GENERAL PRINCIPLES.

sometimes be of more benefit than the actual preparation of defenses.

In the present advanced state of efficiency of fire-arms, artificial cover is, however, of greater importance than ever before. Constructed in the right place, at the proper time, field fortifications may render indispensable service, while their neglect may insure defeat.

5.—While formerly it was the special province of the Engineer to lay out and supervise the construction of defensive works, it has now, under the changed condition of warfare, become the work of the Line as well, and it may be laid down as an accepted rule that the defensive arrangements for a given position are to be made by the troops which are to occupy it:

These changes have affected the art in many ways. The field works now constructed are simpler, ruder, less regular, and less angular than before. An army in presence of an enemy always fortifies, whether in camp, in bivouac, or in line.

6.—Rapidity of execution renders necessary the adoption of fixed types of works in the exercises in time of peace; but these types will sometimes be susceptible of modification in their real application. However, even in war, the endeavor should be to approximate to the regulation forms, for they are deduced from experience and observation, and realize, as well as possible, for each particular case, the best conditions of resistance compatible with rapidity of execution.

The advantage of regulation types is understood at once when it is borne in mind that, upon the battle-field, there should be no hesitation; everyone should stick to his individual role in order to unite efficiently in combined action.

Thorough study and frequent practical exercises, conducted methodically, are indispensable in order to escape feeling one's way, with the loss of time that an insufficient instruction renders inevitable. Upon the battle-field a few minutes may decide the fate of armies in each other's sight.

7.—Fortification, which at first glance may appear to dominate, as representing the "security" and "effectiveness" of an army, the other and apparently less important subjects relating to health and comfort, is, however, so intimately connected with them that neglect of one may render all the others useless. Thus, "bridges," "roads," and "railroads" may, under certain conditions, relate particularly to the effectiveness and security of an army, in connection with Fortification, while under other circumstances they may be as important as various "camping expedients" in the attainment of "health" and "comfort."

CHAPTER II. Fire, Projectiles and Penetration.

8.—Fire as regards its direction is classified as follows:

(1) **Frontal**, when it is delivered at right angles to the front of the enemy's line, and sometimes so termed when delivered straight to its own front.

(2) **Oblique**, when the direction of the fire is at an oblique angle to the front of the enemy's line.

(3) **Enfilade**, which is delivered from positions on the prolongation of the enemy's line. In this case, the line of fire sweeps the enemy's front. When fire is used to sweep along the front of a defensive line and thus enfilade the assailants as they approach the position, it is known as flanking fire.

(4) **Reverse**, when delivered so as to strike troops or lines of defense from the rear.

(5) **Cross**, when the lines of fire from different positions cross on or in front of the enemy's line.

As regards its trajectory it is classified as

(1) **Direct**, when delivered at seen objects at moderate angles of elevation—in the case of artillery when delivered at seen objects, with service charges at elevations not exceeding 15°

(2) **Indirect or Curved**, when delivered with small arms against an unseen object protected by a seen covering obstacle — in the case of artillery, as above, or, with guns, howitzers or mortars with reduced charges at angles not exceeding 15°. Thus firing over an intervening hill at troops sheltered behind it would be an example of indirect fire.

(3) **High Angle**, when used at angles exceeding 15°.

(4) **Grazing**, when the projectile travels approximately parallel to the ground.

9.—The **Artillery Projectiles** used in the U. S. Army are *shell, shrapnel* and *canister*.

12   FIRE, PROJECTILES AND PENETRATION.

Shell. Shell may be classified as common shell and torpedo shell. The common shell is "a hollow cast-iron or steel cylinder with an ogival head closed at one end and filled with powder." The torpedo shell is filled with gun-cotton, or other high explosive. Either shell may be characterized as a flying mine, the chief object of which is to destroy material objects at a distance, though the common shell may also be effectively used against troops.

10.—**Shrapnel** differs from common shell in being filled with bullets, and having only a sufficient bursting charge to rupture the envelope and release the bullets, which then move with a velocity which the projectile had at the moment of bursting. The bullets are assembled in circular layers and held in position by "separators," which are short cast-iron cylinders with hemispherical cavities into which the bullets fit. The shrapnel for the 3.2 inch gun contain 162 bullets ½ in. in diameter, and weighing 41 to the lb. The total number of bullets and individual pieces in the shrapnel is 201.

11. **Canister,** which is practically obsolete, is made of sheet-iron or tin in the shape of an ordinary can, and is filled with bullets held in place by filling the interstices between the bullets with saw-dust, sulphur or rosin; the can is ruptured and its contents dispersed by the discharge of the piece.

12.—The charges in the shell and shrapnel are exploded by means of a combination fuse; by combination fuze is meant one that may be arranged to explode the charge either on impact, by percussion, or at a given time by certain arrangement of the parts of the fuse.

13.—**Field Guns** range up to 6000 yds. but will be seldom used at a range greater than 2500 yds.

14.—**The U. S. Magazine Rifle,** when used as a single loader, has fired 21 aimed shots in one minute, and when used as a magazine rifle, 23 shots in one minute; its range is over 3000 yds. and it is sighted to 1900 yds.

The average heights over which fire may be delivered, are as follows: Man standing 4 ft. 4 in.; kneeling 3 ft.; lying down 1 ft.; field guns 3 ft.

15.—The following thickness of material may be considered as proof against small arm projectiles at all ranges:

Sand..30 in.
Boggy or turfy ground..................................60 in.
Gabion filled with earth.............................. 1
Sand bag well packed, header.......................... 1
" " " " stretcher........................ 2
Packed snow... 6 ft.
Soft wood...40 in.
Oak or other hard wood................................24 in.
Grain sheaves piled...................................16 ft.
Iron plate..$\tfrac{7}{8}$ in.
Steel plate...$\tfrac{3}{8}$ in.
Masonry...20 in.
   Against field artillery.
Earth ....10 to 13 ft.
Snow well packed......................................27 ft.
Masonry (for a short time)............................40 in.

CHAPTER III. Field Geometry.

16.—Before proceeding to that portion of field engineering which involves geometry some of its simplest applications will be explained.

17.—Slopes. The usual description of a slope is by a fraction, the numerator being the height and the denominator the base. Thus, in Pl. 1, Fig. 1, the vertical height is 1-6th part of the base, and the slope is read as 1 on 6. In Fig. 2., the slope is 6 on 1.

18.— To lay out a Right Angle: –First Method. Let A be a point in the line BC, Fig. 3. Lay off from A the equal distances AD and AE. With a radius greater than AD, and with D and E as centers, describe arcs cutting each other at X. Join X with A. Then is XA perpendicular to BC.

Second Method. Find a point such that the distances are in the proportions of 3, 4 and 5: then will the angle included between the two shorter sides be a right angle. Thus (Fig. 4.) with chain or tape measure the distance AD equal to 4 yds. Place one end of tape at D, the other at A, pulling it out and making XD equal to 5 yards, XA equal to 3 yards.

Third Method. At extremity of line, as A (Fig. 4.), assume any point as C. Measure distance CA, set a stake on line BA at a distance from C equal to CA, as D. Set a third stake on line CD at X making CX equal to CD. Then will XA be perpendicular to BA.

19. To erect a perpendicular to a line from a point without. Let X (Fig. 5) be the point without, then, with X as a center, and a distance greater than XA as radius, describe an arc cutting BC at D and E. With D and E as centers, and with a radius greater than DA, describe arcs cutting each other at Y. Join X and Y. Then will XY be perpendicular to BC.

20. To bisect a given angle. Let ABC (Fig. 6.) be the

PLATE 1.

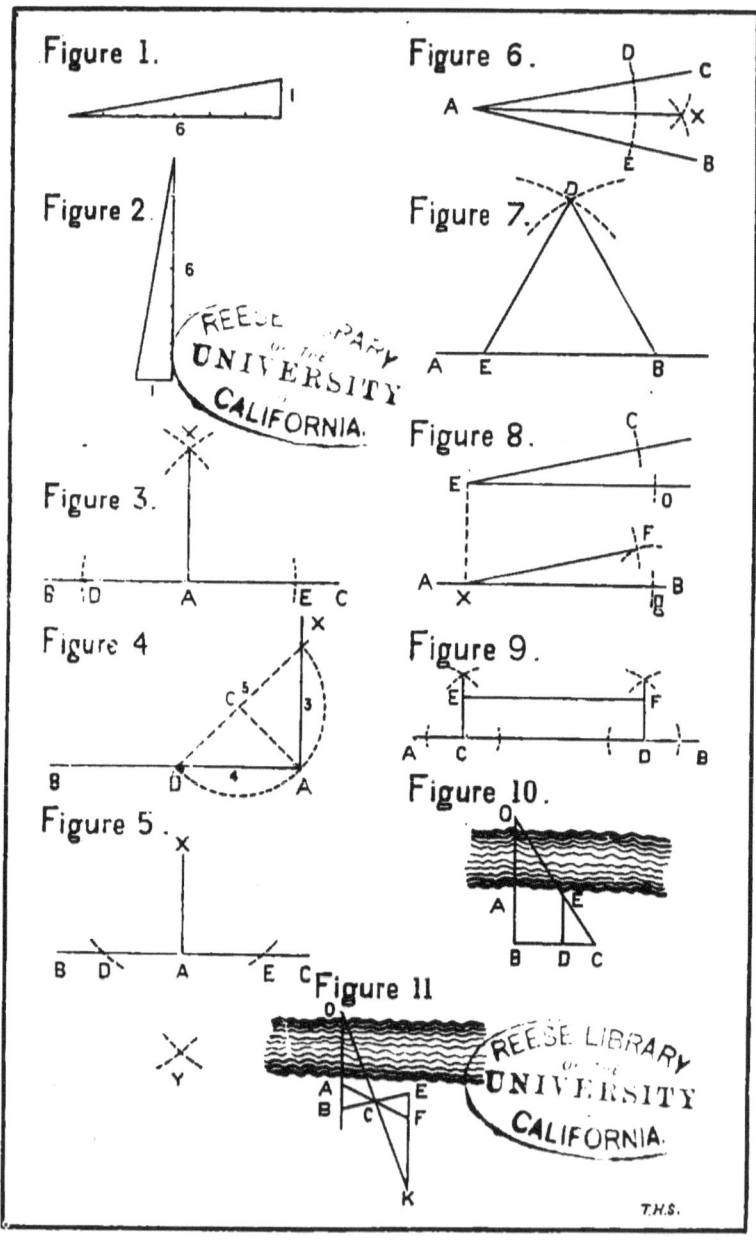

angle. With A as a center, and with any convenient radius, as AD, describe an arc cutting AB and AC at E and D. With D and E as centers, describe arcs cutting each other at X. Join X with A. The line XA bisects the angle ABC.

21.—To lay out an equilateral triangle constructing adjacent angles of 60° and 120°. Let AB (Fig. 7) be a given line. Lay off from B any convenient distance, as BE. Then, with B and E as centers, and a radius equal to BE, describe arcs cutting each other at D. Join D with E and B. The angles DEB, DBE and EDB, are each equal to 60°. The angle AED is equal to 120°. Combining this method with that of slopes an angle of almost any number of degrees can be laid out.

22. To lay out an angle equal to a given angle. Let X (Fig. 8) be a point in the line AB, from which it is required to lay out an angle equal to OEC. Fix the points O and C at convenient distances from E. From X lay off XG equal to OE. Then, with X and G as centers, and EC and OC as radii respectively, strike arcs intersecting at F. Join X and F. The angle FXG is equal to the angle CEO.

23.—To draw a line parallel to a given line and at a given distance from it. Let AB (Fig. 9) be the given line. From any two points, as C and D, erect perpendiculars. On these lay off the required distance CE and DF. Join E and F.

24.—To find the distance between any two points when it cannot be measured directly. First Method. To find AO, take a point B in line with AO and from this point (Fig. 10) lay off any convenient angle, as ABC. At D make EDC equal to ABC. Measure BC, DC and DE, putting E in the line CO. From similar triangles

$$BO : BC :: DE : DC \therefore BO = \frac{BC \times DE}{DC}$$

From the result thus found, substract the distance AB. The remainder is the distance AO.

Second Method. (Fig. 11.) Mark B in prolongation of the line AO. Assume any point as C. Lay off AF, making AC equal to CF: also BE, making BC equal to CE. Prolong EF until a point K is found in line with CO. Measure FK. This is the required distance.

25.—Areas. To find the area of a rectangle. Multiply the base by the height.

To find the area of a trapezoid. Multiply the sum of the two parallel sides by the perpendicular distance between them and take half the product.

To find the area of a triangle. Multiply the base by the altitude and take half the product. Or,

$$\text{Area} = \sqrt{s\,(s-a)\,(s-b)\,(s-c)}$$

in which s is the half sum of the three sides a, b, and c. Or,

$$\text{Area} = \tfrac{1}{2}\,a\,b\,\sin C$$

in which a and b are two sides and c the included angle.

26.—The Field Level (Pl. 2, Fig. 1) consists of three strips of wood A, B and C, each ½ in. thick and 2 in. wide. A being 62 in. long, B and C each 44.42 in. The distance between centres on A is 60 in. on B and C 42.42 in. This makes a right angle between B and C. There is a thumb nut at E clamping the arm B to the arm A when the level is used. The screw at F projects, holding the arm B, when folded, up. There is a stud at H, affording an attachment for a plumb bob. There are permanent joints between B and C, and A and C.

Fig. 1 shows the level and its joints, plumb bob for reading slopes, and spirit level. Fig. 2 shows side for protracting angles.

27.—Uses of Level. The level may be used as follows:

(1) As a spirit-level, the level being on the edge C.
(2) As a square for setting out a right rngle.
(3) As a protractor.
(4) For setting off slopes.
(5) As a mason's level with a plumb bob.

PLATE 2.

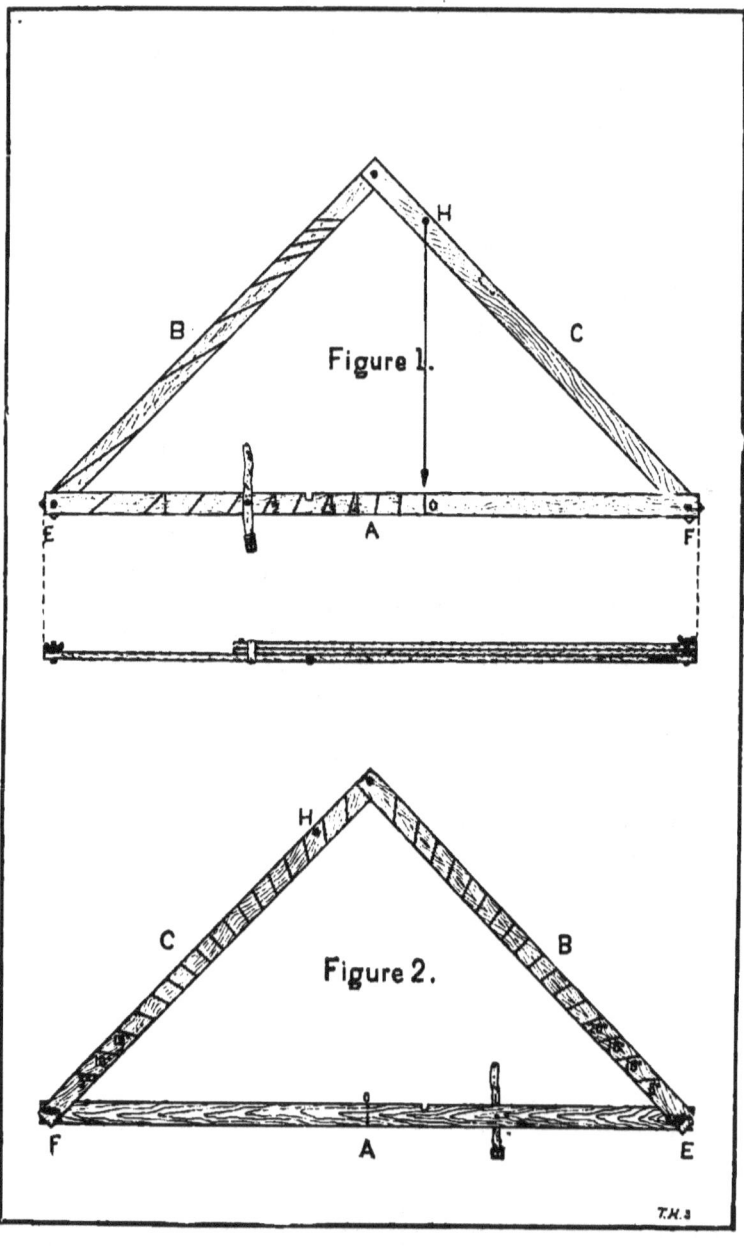

PLATE 3.

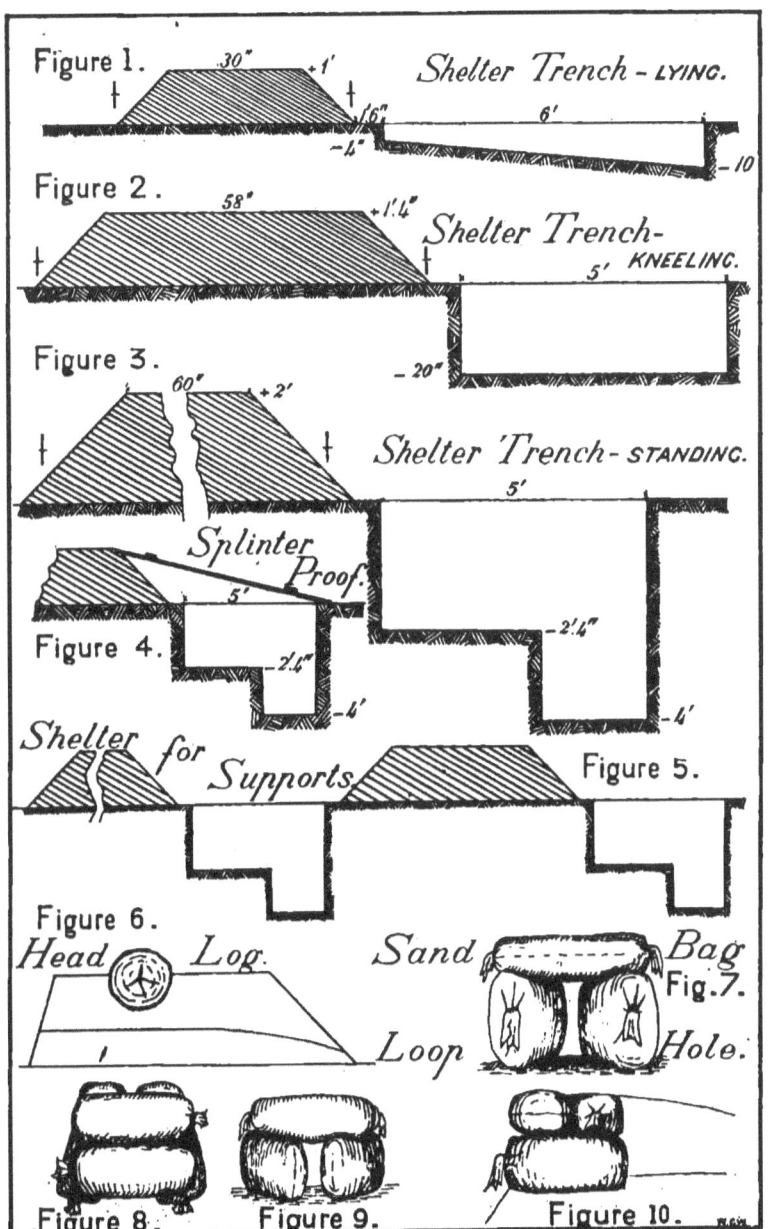

CHAPTER IV.—Hasty Intrenchments, Gun Pits and Epaulements.

28.—The intensity of fire made possible by the fire-arms of to-day renders some form of shelter on the field of battle imperative. Circumstances may occur when advancing lines of skirmishers will find natural shelter, but in most cases artificial cover must be constructed on the spot. The fortifications used on the field of battle depend, as to their position, extent, and use, on the ground and on the tactical advantages to be gained, and, in conformity to this idea, they are constructed at the time of the battle and not before. They are called "Battle" or "Hasty" intrenchments and consist of cover for

(1) Skirmishers lying, kneeling, or sitting.

(2) Firing line, Supports and Reserves kneeling, sitting, or standing.

(3) Gun pits and Epaulements.

29.—Pl. 3, Fig. 1, shows the section of shelter trench for skirmishers—lying down. It gives earth protection to the depth of 30 inches, which will stop small arms projectiles, except when struck in the most favorable spot for penetration. The time with one large spade and one pick—one man—about 15 minutes; with small spade, about 20 to 25 minutes.

30.—For men kneeling or sitting in two ranks, cover is gained by deepening the trench already dug to 20 ins. and making it 5 ft. wide, thus obtaining a trench having a defense of 1 ft. 4 in. high and about 58 in. thick. Time about 30 minutes—small spade. If single rank is used the trench need only be 2 ft. 6 in. wide. (Fig. 2.)

31.—Cover standing is obtained by deepening the kneeling trench to 48 in., leaving a step 20 in. high and 3 ft. wide next the front wall of trench so as to facilitate the leaving of the trenches to the front and also to serve as a banquette for men firing over defense. The earth defense should thus be 60 in. thick and 2 ft. high; time—infantry spade—1½ to 3 hours. Whenever it is possible the interior slope should be as near vertical as possible. (Fig. 3.)

32.—When isolated trenches or rifle pits are desired for the use of riflemen they should be made with the same section as the trenches for cover standing, 2 paces being taken as the length of trench for each man. Should a more thorough shelter be desired than is given by the cover standing the earth defense may be increased by digging a ditch in front, care being taken that it is so far to the front that sufficient distance is given to develop the existing works into the normal field works—*i. e.*, 15 to 18 feet.

33.—The trenches here illustrated are all made on level ground and are simply types showing the best form and giving general ideas as to the time required to construct cover. They must be varied according to the kind of earth in which they are to be made and such modifications introduced as will strengthen the cover without increasing the time used. The location of trenches depends primarily on tactical considerations, secondarily on the nature of the ground. Sand or stony ground should, if possible, be avoided in selecting the position of trenches. The eye should be placed a distance above the ground equal to the height of the completed parapet in order to locate the trench so as to obtain a clear field of fire to the front. Shelter trenches should always occupy a position giving the greatest development of fire to the front and are generally constructed at the crest of the most abrupt slope.

34.—In all trenches, except those for skirmishers lying down, intervals in the line should be left for the passage to the front of Artillery and Cavalry—this is especially necessary when the cover standing is used. These intervals will vary in width according to circumstances but should never be so wide as to preclude their defense by the trenches adjoining the opening. When stones are encountered in digging they should be imbedded in the parapet and well covered with earth, as they splinter badly when struck by bullets or shrapnel fragments.

35.—When men are required to remain in the trenches for any considerable period they should be provided with splinter proofs or shelters of some kind. Planks, old lumber, doors, etc., may be used, in absence of which small poles are available; they should be laid with one end on the parapet and the other resting on the earth in rear of the trench and then covered with 2 to 3 inches of earth. This defense, while not proof against bursting shells, will protect the men from dropping bullets and shrapnel fragments. (Fig. 4.)

36.—Loop Holes may be provided by placing head logs half imbedded in the parapet or resting on sandbags on the crest and leaving spaces beneath for the rifle, or with sandbags, two being laid lengthways to form the sides and two resting on top, as shown in Figs. 6, 7, 8, 9, 10. Brushwood may be used by laying it on top of the excavated loop hole and covering it over with earth; all loop holes for musketry splay inwards.

The best practice is not to use head logs or construct loop holes (unless the force is acting solely on the defensive), as their use impels men to hesitate to leave their cover when the advance is ordered.

37.—When necessary to intrench supports or reserves, the cover kneeling or standing should be used in parallel rows and close to one another. Fig. 5.

38.—The *advantages* of the shelters for men lying and kneeling are,

(1) That they present but small difficulties to the advance of Cavalry or Artillery over and through them.

(2) They offer but a small target to artillery fire.

(3) They are not often penetrated by rifle bullets.

(4) They are easily surmounted by their defenders when taking the offensive.

Their *disadvantages* are,

(1) That, being of low camp'n and relief, they often have their field of fire limited by small folds of the ground, although this may be eliminated to a large extent by care in selecting the site.

(2) In wet weather they become muddy and uncomfortable. In all intrenchments it is well to loosen the earth in front, to lessen the effect of shells striking them. It is also necessary to disguise the intrenchment by branches, sods, etc., so as to deceive the enemy in regard to their location.

39.— Cover for guns and caissons is obtained in either of two ways: –

(1) By sinking the guns below the surface of the ground and building a parapet with the excavated earth. *Called Gun Pits.*

(2) By building an epaulement in front of and around the gun, the gun in this case resting on the natural surface of the ground. *Called Gun Epaulements.*

The Gun Epaulements shown (Pl. 4, Fig. 5) may be constructed while the gun is in action, which cannot be done with any form of pit, as pits must be completed before the gun can be put into action. Gun pits of various kinds are shown in Figs. 1, 2, 3 and 4.* When the ground is soft, or in plowed fields, the pit is the better construction for use.

40.—In constructing any earthworks plows may be used to advantage to loosen up the earth. The furrows should begin at the back line of the trench and then return along the front line until the earth is loosened in all the trenches; two or three plows following each other at intervals can be used with advantage.

---
* Figs. 1, 2 and 4, Pl. 4, are from U. S. Artillery Drill Regulations.

PLATE 4.

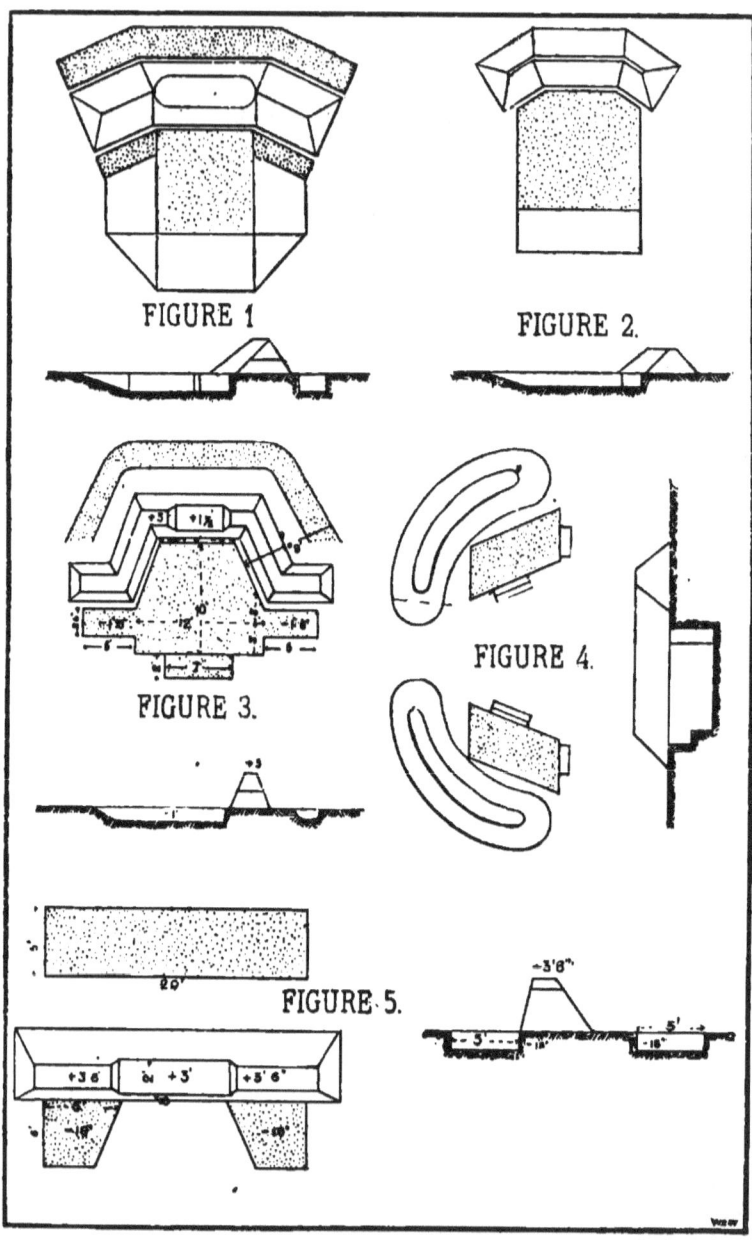

PLATE 5.

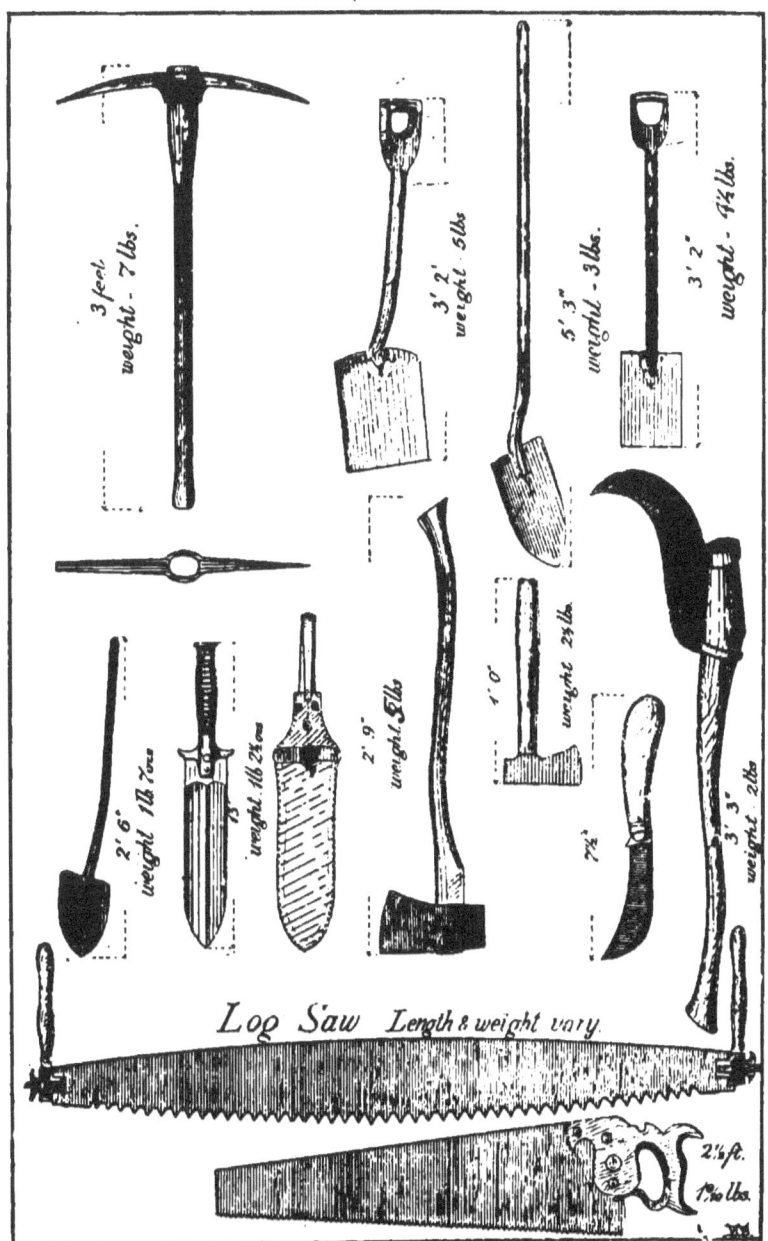

CHAPTER V.—Clearing the Ground.

41.—The tools more especially used in the field may be divided into two classes.

(1) **Intrenching tools,** such as the pick, the shovel (long and short handled), the spade, the picket shovel, and the hunting knife (Infantry equipment.)

(2) **Cutting tools,** such as the ax, the hand ax, the log saw, the hand saw, the linked felling saw, the gabion knife (pruning knife), the hunting knife, the bush hook and the wire cutter. (Pl. 5 and 6.)

42. The choice of a defensive position in which the foreground is free from obstructions and favorable to the defenders' fire is of the utmost importance: more or less clearing, however, will usually be necessary. Clearing must be systematically done and, as in all other work, should be undertaken by complete organizations or parts of organizations, under their own officers.

43.—The extent (theoretical) to which the foreground should be cleared is equal to the effective range of the defenders' weapons. Practically, as wide a space within this limit, is to be cleared, as is consistent with the time and labor available. Brushwood and standing timber most often screen the enemy's advance and steps should be taken to remove them.

44.—The tools usually employed in felling heavy timber are the ax and the log saw, the former being the most common, although inexperienced men acquire familiarity with the latter more quickly. When using the ax the cut should be commenced on the side toward which it is desired the tree should fall, ropes being used to incline it in that direction, if necessary; if immaterial which way the tree falls, then attack it on the side toward which it leans; after cutting it a little more than half through change over to the other side and commencing about six inches higher up, cut until it falls. In using the saw it may be necessary to wedge the cut or use other means in order to keep the saw free: the teeth should be set wide.

Both saw and ax may be used, in which case the ax should be used on the side toward which the tree is to fall and the saw on the other side. (Pl. 6.)

45.—**Trees** would ordinarily be cut within a foot of the ground because a greater height would afford cover. A man should cut down a hard wood tree 1 ft. in diameter in 10 minutes and one of soft wood of the same size in one third the time. The hand ax, hand saw, and hunting knife are useful in felling small trees, ropes being attached to bend them, and the cut being made on the convex side.

Felled timber must be removed, if of such a size as to afford shelter to the enemy. It is utilized in making field casemates, magazines, etc.

46.—**Brushwood** can be cleared at the rate of about 12 sq. yards per man per hour. The men should be extended at about 4 paces interval, using the bush hook, hatchet, or hand ax, together with the gabion or hunting knife.

47.—**Grain, grass,** or **weeds** must be trampled by men in line, mowed, or burnt.

48.—**Hedges, fences,** and **walls,** if not perpendicular to the front, must be removed. Live hedges should be pulled to one side in order to give the axmen greater freedom. (Pl. 6.) Fences can ordinarilly be demolished with axes, walls by battering them down or blowing them down with explosives. (Chapter XX.)

Buildings may be battered down, burned, or demolished by explosives according to circumstances. In the case of buildings and walls it will usually be necessary to remove the debris, which can be used for filling hollows.

PLATE 6.

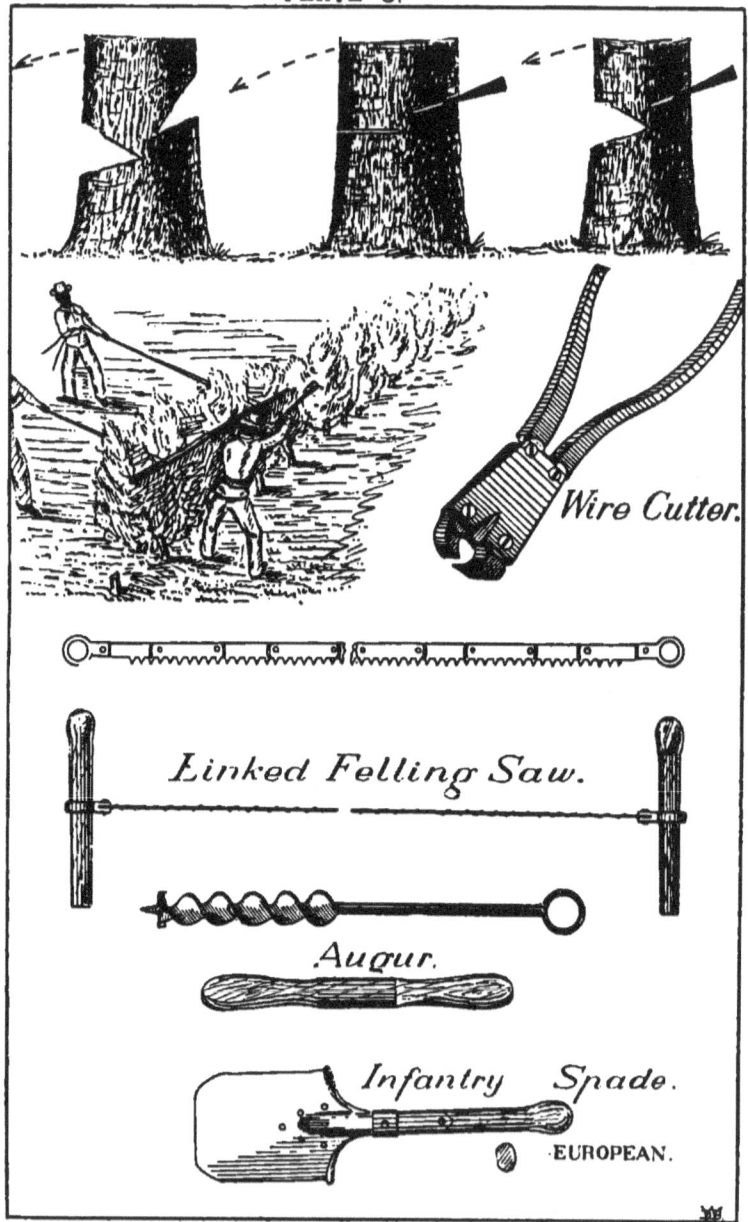

PLATE 7.

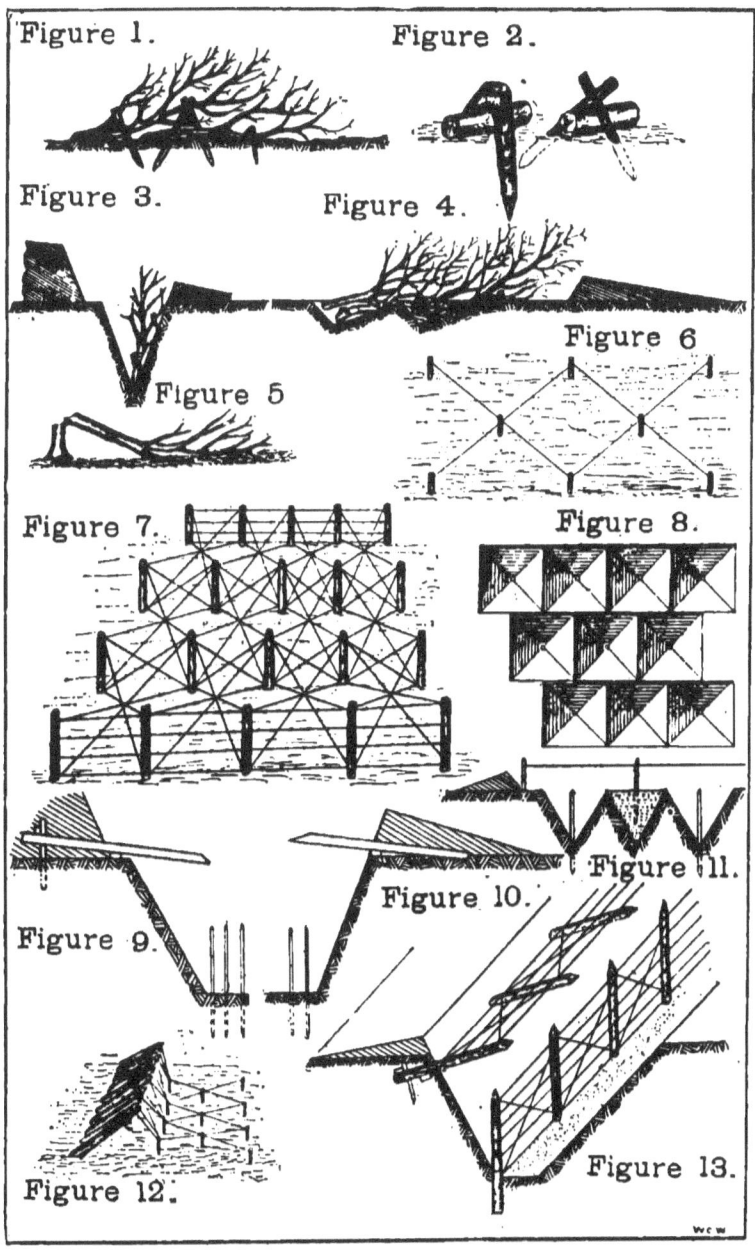

CHAPTER VI.—Obstacles.

49.—Obstacles have for their object the holding of the enemy under fire while checking his advance and breaking up his formation.

(1) They must be within the effective zone of the defenders' fire and must be so arranged as to offer the least obstacle possible to an advance from the side of the defense.

(2) They must be concealed as far as possible from the view of the assaulting party, so that they may come upon them as a surprise.

(3) They must be difficult of removal under fire, and, if possible, should be of such construction as will necessitate the use of tools not usually carried by troops.

(4) They should, if possible, be so placed as to be secure from the fire of the enemy's artillery, and so constructed that, if struck by his projectiles, they will suffer small damage.

(5) They must offer no shelter to the enemy.

50.—Abatis, on account of the ease with which it can be constructed, is the obstacle most used.

It consists of branches of trees about 15 feet long, laid on the ground, butts pointing to the rear, all small twigs being cut off, and all large branches pointed and interlaced. The abatis should be 5 feet high.

The branches are secured to the ground by forks, wire, or by logs laid over the butts of the branches. The use of wire to hold down the branches is recommended, and when used should be also passed from branch to branch so as to form an extra form of entanglement. When more than one row of abatis is used the branches of one row overlap the butts of the next one in front. (Pl. 7, Figs. 1 and 2.)

The abatis most easy of construction is that made by felling trees towards the enemy in such manner as to leave the fallen part still attached to the stump, the branches are then pointed as described before. (Fig. 5.)

51.—Abatis is often placed in the front of works when the ditch is so shallow as to present little or no obstacle to an assault. When so used they are placed upright and well tamped in. In all cases, especially when small branches are used, it is better to sink the butts in triangular pits, and, when the branches are in place, fill in with earth and tamp well. (Figs. 3 and 4.)

In all cases where exposed to artillery fire a glacis should be constructed in front of an abatis, so as to protect it from injury.

52.—Low Wire entanglements are formed by driving into the ground stakes about 18 in. long. The stakes should be driven in rows about 6 feet apart, the stakes in each row being opposite intervals in adjacent rows. The heads of the stakes are connected by stout wire wound around them. To make this more effective, do not clear the ground, but allow bushes, brush, etc., to remain in place. (Fig. 6.) Use 1 ft. of wire to 1 sq. foot of ground covered.

53.—High Wire entanglements are constructed in the same manner, except that the stakes should be at least 4 ft. high, and placed 6 to 8 feet apart. The head of each stake is connected with the foot of the stake diagonally opposite, the line of posts in front and rear being finished off as fence panels with barbed wire. The use of barbed wire is not advised for the interior crossed work on account of the danger and difficulty in working with it.

Roughly, one yard of wire is necessary for each square foot of entanglement. Ten men can make about 9 square yards of this entanglement in one hour. This work does not require trained men. Wire entanglement, either high or low, is useful on the glacis of field works, as it holds the attacker under fire at the most favorable point." (Fig. 7.)

54.—Palisades consist of rows of trunks of trees or of squared trunks, 8 or 10 feet high, planted close together and pointed on top. When material is at hand, ribband pieces should be spiked on the inside along their tops about a foot or two below the points to hold them steady. They are used to advantage in the bottoms of ditches or to close the gorge of field works. (Pl. 8, Fig. 1.)

PLATE 8.

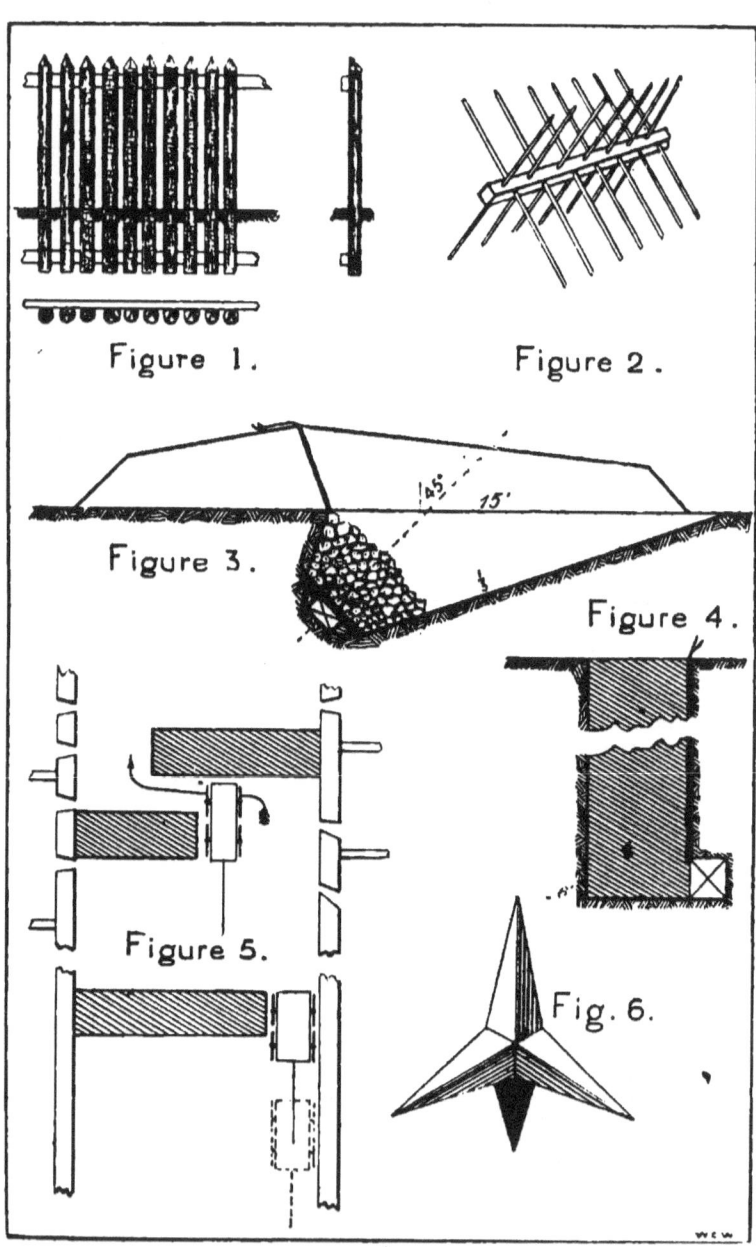

OBSTACLES. 41

55.—**Fraises** are palisades arranged horizontally or much inclined and are much used at the foot of the exterior slope and at the top of the counterscarp; in the first position they point down and in the second upward. In each case, the ribband or strip is spiked on and laid against the ground near the edge of scarp or counterscarp, as the case may be, another one being spiked to the inner end of the fraise, thus the outer ones give good bearing surfaces and do not break up the crest, and the inner one gives a bearing for staking and tying down. The slopes described above are given so that unexploded shell will always roll away from the parapet. (Pl. 7, Figs. 9, 10, 12 and 13.)

Fraises may with advantage be made of barbed wire in the form shown, care being taken that all wire when finished is on top of the wooden supports. The advantage of this variety of fraise is that it is little damaged by artillery fire and is very difficult of removal.

When time is pressing fraises may be made of branches of trees with the butts well sunken and staked down.

56. —*Crows' feet, Chevaux-de-frise,* and *planks full of spikes* have been used in the past as obstacles to an advance, but the two former are not now issued for use in our service, and the latter is one not easily made in the field. Such obstacles require much time and material in their construction and are not treated of here, as they do not fall properly in the domain of Field Engineering; their value in any event is not commensurate with the difficulty of their preparation. (Pl. 8, Figs. 2 and 6.)

57.—**Small Pits** are square on the top, 3 feet on a side, and are pyramidal in shape; they are 2 ft. 6 in. deep, and have a pointed picket driven in the center of each.

In digging these pits a glacis should be formed in front of the row nearest the enemy and, to avoid filling the pits with earth thrown from the others, the row farthest from the glacis should be commenced first. One man can make 10 pits per day in easy soil. (Pl. 7, Figs. 8 and 11.)

Small pits may be surmounted by a low wire entanglement, making a very serious obstacle.

58.—**Fords** may be made impassable by strewing them with harrows, points up.

59.—**A Fougass** is a mine so arranged that upon explosion a large mass of stones or shells are projected towards the enemy. (Pl. 8, Fig. 3.)

To make a fougass, dig a hole in the shape of a frustum of a cone, inclining the axis in the direction of the enemy, so as to make an angle with the horizon of about 45 degrees. The sides should splay outwards at an inclination of 12 degrees from the axis. The powder charge is placed in the bottom of the hole — preferably in a box—and in front of this a platform of wood about three inches in thickness: on this are piled stones, brick, etc. The mine is exploded by means of electricity or common fuse. Care must be taken in digging the hole for the fougass that the line of least resistance is in all cases in the axis of the hole, *to be sure of this, throw the excavated earth upon the crest towards the defenders' side* and ram well, allowing earth to enclose the sides of the excavation in the manner shown in cut.

Fougasses are useful in defending boat-landings, roads, etc.

The following empirical formula may be taken for determining the charge of powder for fougasses: $P = \frac{s}{150}$, in which P and s represent the weight in pounds of the powder and stone.

When broken up, a cubic foot of limestone weighs 96 lbs.

60.—**Land Mines** are small mines placed in the line of advance of the enemy and exploded either by electricity or fuse from the defense. The small mines are made by digging holes from 2 to 3 yards deep, placing the charge in a box in a recess excavated in one side of the hole, and refilling with the excavated earth, tamping well. The wires are carried back in a small trench to the work. In common earth, the charge for 2 yards deep is about 25 lbs., and for 3 yards deep about 80 lbs.; the diameter of the crater formed will be about twice the depth of charge. (Fig. 4.)

61.—**Barricades** are used to prevent the passage of the enemy through roads, streets, and defiles generally.

They may be made of any material at hand, paving stones, overturned carts, barrels filled with earth, stones, and articles of like nature. They should be built so that a passage is always left for the defenders, but means should be at hand to close the opening quickly—a wagon may be used for this purpose, being drawn away from the opening when a passage is desired.

The houses on either side should be loop-holed and used to flank the defense. Overturned wagons, broken furniture and debris from the adjacent houses make a very good *obstacle* and should be placed in front of the barricade to ward off cavalry charges. (Fig. 5.)

PLATE. 9.

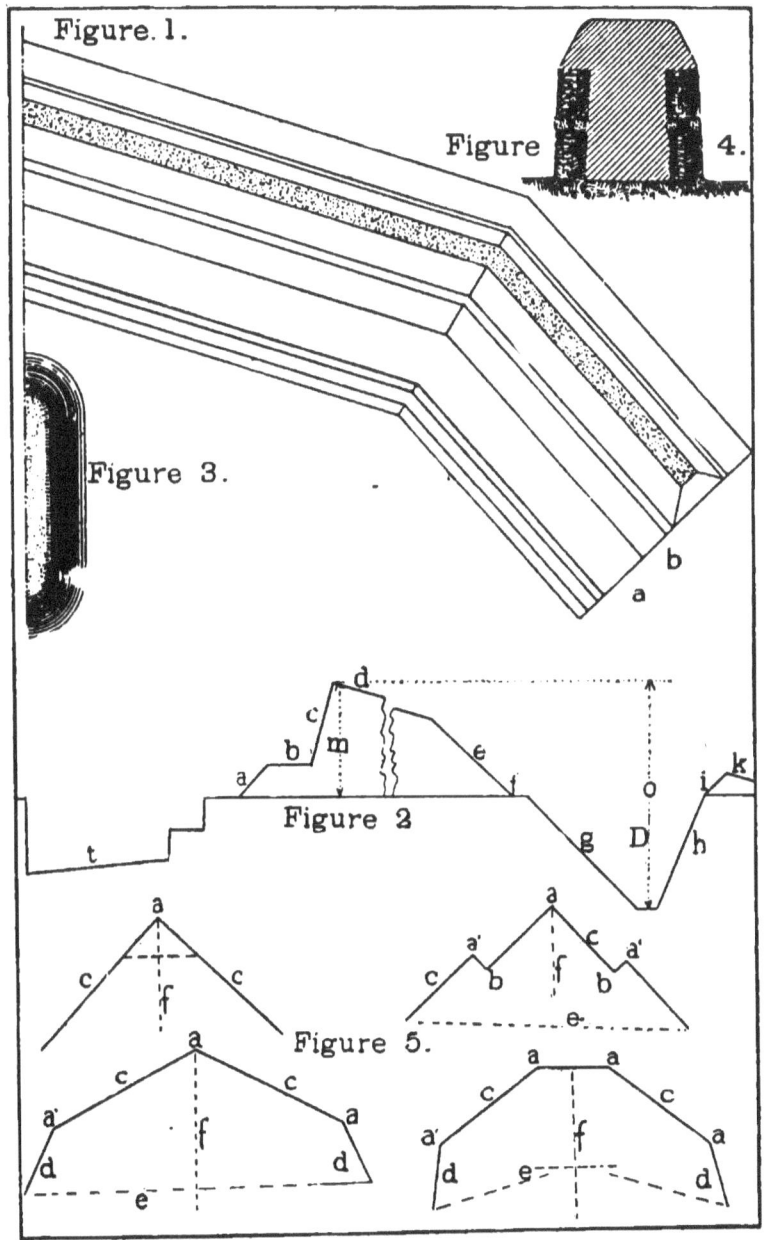

CHAPTER VII.—Field Works.

62. When a position is to be held for a considerable period and when time is available, more deliberate defenses than the Hasty or Battle Intrenchments (Chapter IV) are constructed. These are known as *Field Works* and usually require a minimum of 6 hours for construction. The conditions to be fulfilled, besides those necessary for a defensive position (Chapter I), are

(1) That they must afford protection against both rifle and artillery fire.

(2) That they must be of suitable size for the garrison that is to occupy them.

(3) That they should have suitably constructed casemates to shelter the garrison at night.

Field works may be constructed for the defense of a single object, as a bridge, a ford, etc., or they may occupy the key points in a long line of defense, in which case they should be located so as to afford mutual protection, the intervening space either being left open or occupied by shelter trenches.

Before proceeding to the study of Field Works, a brief synopsis of the technical terms used in connection with them will be necessary.

63.—A **Parapet** is a bank of earth thrown up to cover the defenders while firing.

64.—The **Trace** of a work is its outline in plan: the term is often applied to the horizontal projection of its interior crest. (Pl. 9, Fig. 1.)

65.—The **Profile** is a cross-section of the work made by a plane perpendicular to the interior crest. (Fig. 2.)

In the profile, the various parts are named as follows:

a. Banquette slope. e. Exterior slope. d. Ditch.
b. Banquette tread. f. Berm. i. Interior slope of glacis.
c. Interior slope. g. Escarp. k. Glacis.
d. Superior slope. h. Counterscarp. t. Trench.

66. **The Interior Crest** is the intersection of the Interior and Superior slopes: sometimes called the magistral line. ("a" Fig. 1.)

67. **The Exterior Crest**—that of the Superior and Exterior slopes. ("b" Fig. 1.) The thickness of parapet is the horizontal distance between interior and exterior crests.

68. **A Traverse** is a bank of earth inside a work to protect some portion of it from direct fire. When the protection afforded is from reverse fire, the traverse is sometimes called a **Parados**. (Figs. 3 and 4.)

69. **An Embrasure** is a revetted opening in the parapet, through which field guns may fire. It is said to be Direct or Oblique according to whether its axis is perpendicular or inclined to the line of parapet.

70. **A Gun Bank** ~~or Gun Emplacement~~ is a raised mound, by means of which field guns may fire over the parapet. Guns thus placed are said to be "en barbette."

The relative advantages of Embrasures and Gun Banks are as follows:—

Embrasures afford greater protection to the gunners, but

(a) They afford a very limited field of fire.

(b) They weaken the parapet and require frequent repairs.

(c) The place of the gun when not in action cannot well be used by Infantry.

The conditions as to Gun Banks are the converse in each case.

71. **The Command** of a work is the height of its interior crest above the ground on which it is constructed. ("m" Fig. 2.)

72. **Its Relief** is the height above the bottom of the ditch. ("o" Fig. 2.)

73. **The Plane of Site** is a plane tangent to the ground on which the work is constructed.

74. **The Terreplein** is the surface of the ground inside the work and does not, of necessity, coincide with the plane of site, since the whole interior of the work *i. e.*, the terreplein may be lowered for the purpose of securing more cover.

When the banquette tread is more than two feet above the terreplein, its slope may be stepped with fascines or planks: this has the advantage of giving more interior space but tends to produce confusion on the part of the defenders, especially in a night assault.

75.—**The interior slope** is usually made as steep, up to four on one, as the revetment will stand.

76.—**The superior slope** is necessary in order to secure the best fire effect on the ground immediately in front of the work. It weakens the parapet near the interior crest, however, and this defect increases as the slope is made steeper, hence, it should be as slight as is consistent with good fire effect.

The degree of this slope is regulated by the principle that fire from rifles resting on its surface should not pass more than three feet above the glacis, or, when there is no glacis, above the outer edge of the ditch. It will thus depend on

(1) The command of the work.
(2) The inclination of the plane of site.
(3) The distance from the interior crest to outer edge of ditch.

The slope should not exceed one on four; one on six (normal) is better, and then, if necessary, make a glacis of the requisite height.

77.—**The exterior slope** should be as gentle as two on three, if possible, owing to the fact that steeper slopes are soon destroyed by artillery fire.

78.—**The Berm** may be as great as 6 ft. in width; ordinarily it would not be greater than 2 ft., while in favorable soil none may be left at all.

Advantages of berm.

(1) It relieves the edge of the ditch from the weight of the parapet and thus prevents caving, in loose soil.
(2) It enables the parapet to be thickened.

Disadvantages:

(1) It affords a footing in an assault. (This, however, may be partially remedied by use of obstacles.)

79.—The slope of the escarp and counterscarp should be equal to or greater than the exterior slope, the latter being as steep as the earth will stand.

80. The Glacis should be parallel to the superior slope, in order to get the best fire effect from the crest.

81. If the parapet does not require much earth, and the ditch is required as an obstacle, it may be made triangular in cross section. This form gives the greatest depth and prevents the assailants from forming in the ditch, but it is difficult of construction. Eight feet may be taken as the extreme depth of ditch and twelve feet as the extreme height of parapet. The width of the ditch varies with the amount of earth required—12 ft. at the top being a minimum.

A parapet with trench and ditch affords cover in the shortest time possible; each foot of depth in the trench means 2 ft. of cover, plus the additional protection afforded by the earth from the ditch.

A parapet with ditch alone affords greater cover to the ground in rear and better command of ground in front, but its height makes it more conspicuous.

82. Referring to traces of various works (Pl. 9, Fig. 5.)—

a, is a salient angle.
a', is a shoulder angle.
b, is a re-entrant angle.
c, c. c, are faces.
d, d, d, are flanks.
e, e, is the gorge.
f, is the capital.

83. Field Works are classified with reference to their trace, as

(1) Open, which have thick parapets on exposed sides, the rear or gorge being open.

(2) Closed, in which the thick parapet is continuous.

(3) Half-Closed, which only differ from the "open" in that the gorge is closed by obstacles, stockade work, or shelter trenches.

Advanced works within rifle range of the main defensive line, as well as those in positions where the flanks are secure (as a bridge head), should usually be "open." Works in main line and advanced works beyond rifle range should be "half-closed;" those in isolated positions or on the flanks of a defensive line—"closed."

Open works have the advantage over closed, of affording greater freedom of movement to the defenders, and, in the event of capture, of being exposed to fire and assault from the works in rear.

PLATE 10.

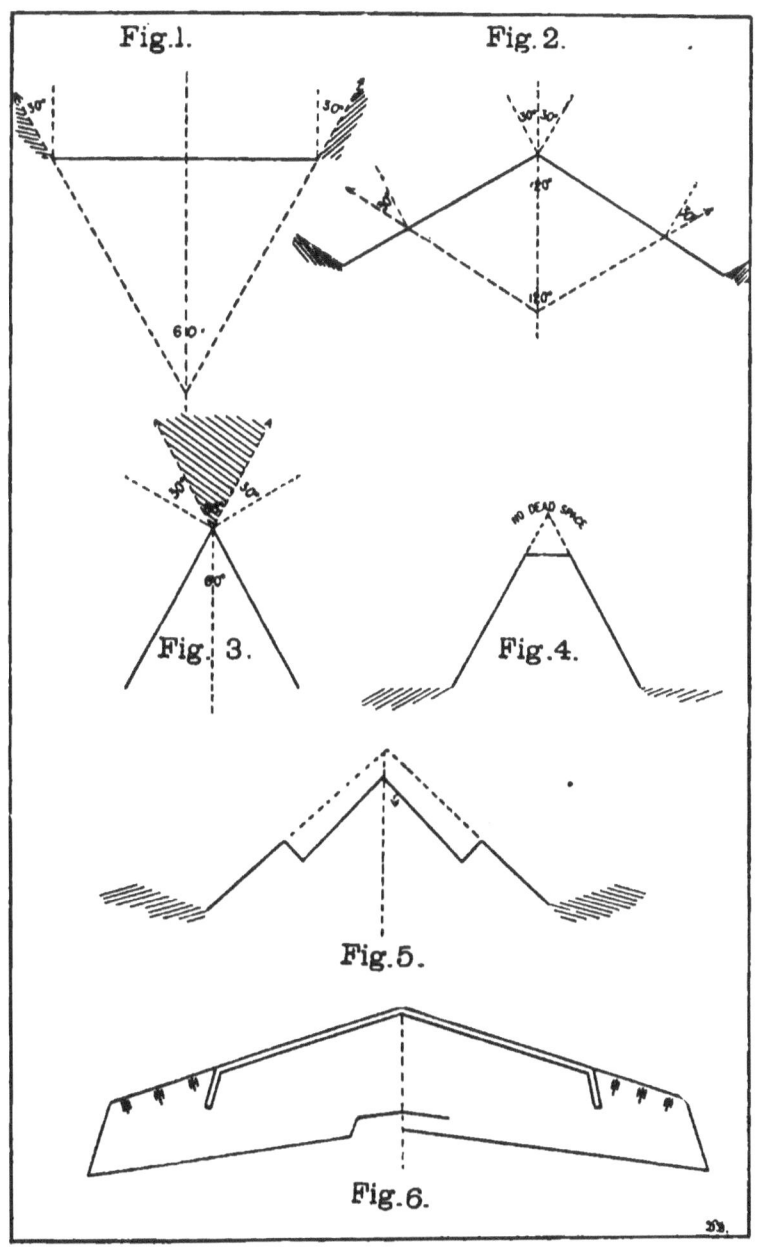

FIELD WORKS. 51

Closed works, while affording greater protection from assault, are liable to have their parapets exposed to enfilade or reverse fire, besides which the available interior space is much reduced.

84.— Forts and Redoubts (Closed Works) are distinguished by the former having re-entering angles, thus affording defense of the ditch from the parapet, both conditions being lacking in redoubts.

Redoubts, as compared to forts, are of simpler trace, do not require so large a garrison, and afford better frontal fire; but, as they have no ditch defense (unless caponiers and counterscarp galleries are constructed), they should be traced to support one another.

85.—With respect to caponiers (Pl. 20) and counterscarp galleries—the former, if sunken, as is usually necessary for protection against artillery fire, may, become untenable in rainy weather; while communication with the latter is difficult and may by the enemy, be rendered impossible. The objections to these forms of ditch defense are so great, and their use so limited, where proper frontal fire and obstacles are possible, that their construction is seldom necessary.

86.—The Sector of Fire is a term used to designate the angular space in front of a work which is swept by its fire (30° on each side of a perpendicular being considered the limit of oblique rifle fire.) Thus, a straight line of parapet has a sector of fire of 60° (Pl. 10, Fig. 1), while, in a redan, it varies with the angle at the salient. With a salient of 120°, the sector of fire is evidently 120° (Fig. 2), with a 60° salient, there will be an undefended space of 60°. (Fig. 3.) This undefended space may be done away with by blunting the redan. (Fig. 4.) A redan with shoulder angles (Fig. 5) furnishes a ditch defense in front of the shoulders and does away with part of the dead space in front of the salient but it is difficult of construction and is not usually resorted to.

87.—For reasons given in Chapter XI, it is often desirable to place the guns outside the work; in which case some plan like Fig. 6, may be adopted, the single line representing a shelter trench.

88.—Defilade of Field Works. In order that Field Works may fulfill the condition of screening the occupants from the fire and view of an enemy, the problem of defilade arises.

This may be defined as the operation of regulating the direction and command of the earth cover so that the interior of the work is protected from the direct fire of an enemy.

The problem resolves itself into two distinct parts—
(1) *Defilading in plan.*
(2) *Defilading in section.*

89.—Defilading in plan. This involves the selection of the trace of the work (its position having been previously chosen.) The trace will vary with the plane of site, the terrain in the immediate vicinity, the proximity of high ground that the enemy may occupy, and the time available for construction. A plane of site sloping to the rear is obviously the easiest to defilade, and one sloping toward the enemy, the most difficult. Salients should occupy commanding ground, the lower portion being taken for the re-entrants or for the gorge. The longer faces of a work should lie in the direction of lower or inaccessible ground, so that they connot be enfiladed.

With commanding ground in front, the work is more difficult to defilade in proportion to its depth, therefore, have longer faces opposed to the high ground and make the work as shallow as is consistent with other conditions.

As a rule, the longer faces of a work must lie so that the defenders can bring as direct a fire as possible in the direction of expected attack.

All the foregoing conditions as to defilading in plan, cannot, in the usual case, be satisfied, but the object to be attained must be kept constantly in view, and, in selecting the trace for a work, an officer's ability will be shown by the skill with which he harmonizes the various diverse requirements.

After the careful selection of the trace, as already indicated, and marking it by pickets, the problem is completed by defilading the proposed work in section.

90.—Defilading in Section. With a horizontal site and only level ground toward the enemy, a constant command of 8 ft. is sufficient to protect the whole interior of the work.

On an irregular site, or when necessary to place a work in a position commanded by higher accessible ground, the necessary protection of 8 ft. may be attained in one of three ways—
(1) By raising a parapet.
(2) By lowering the terreplein.

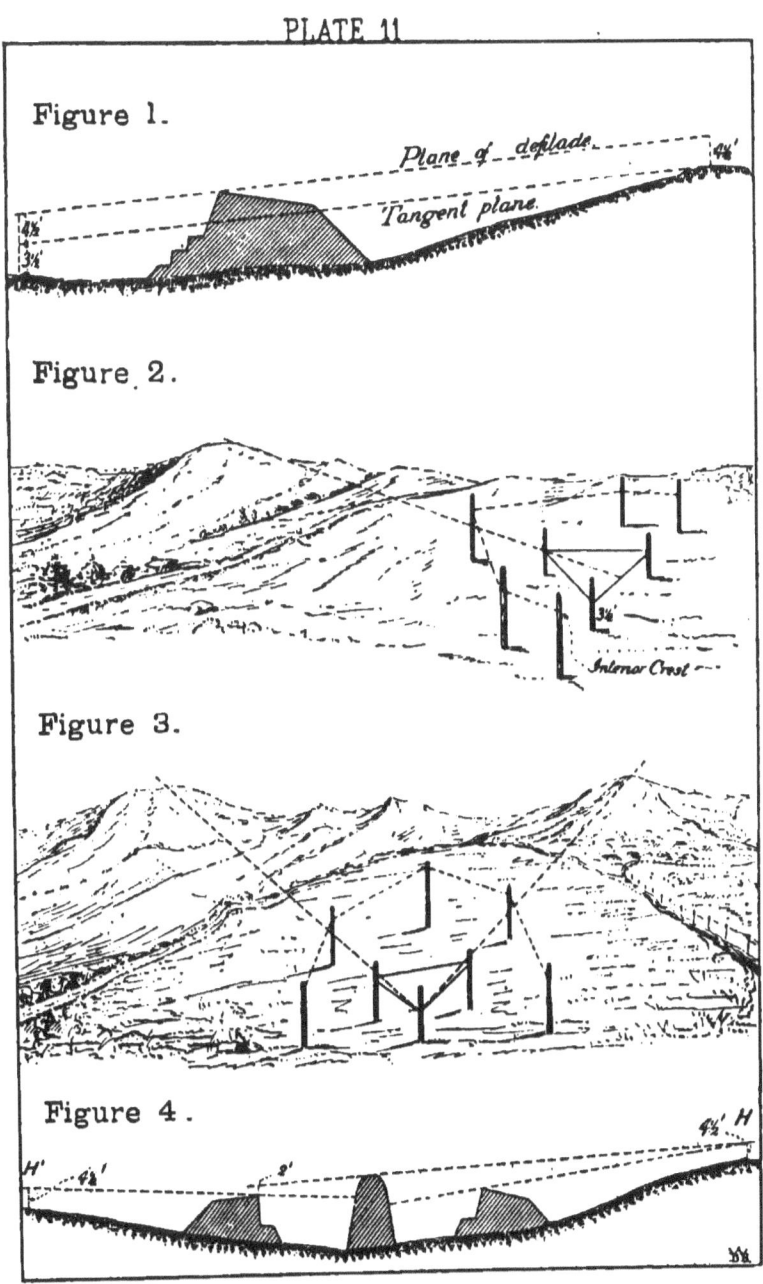

(3) By use of traverses, parados, bonnets, etc.

To determine how much protection is needed, suppose, for example, the proposed work is a lunette. Plant poles at the salients of sufficient length to reach the interior crest of completed work. Place two pickets at the gorge, about 6 ft. apart, one on each side of the capital and a third 8 ft. to the front. Tie a string to the rear pickets, 3.5 ft. from the ground, the string passing round the third stake. (Pl. 11, Fig. 2.) Taking position behind the horizontal string, have an assistant move the string on the forward picket until it comes into the plane fixed by the eye, the horizontal string and the highest point of the dangerous ground. This plane, which is now established by the string triangle, is called the *tangent plane*. A plane parallel to this and 4.5 ft. above it is known as the *plane of defilade*. (Fig 1.) The proper height of parapet at the salient and shoulder angles is now fixed by sawing off the poles 4.5 ft. above the points in which the tangent plane cuts them. This will evidently give 8 ft. cover at the gorge, at which point the height of parapet of the flanks is 8 ft.

If it is found that the required height of parapet exceeds 12 ft., the plane of defilade may be lowered not to exceed 1.5 ft. This will still give 6.5 ft. protection at the gorge.

If this proves insufficient, either traverses must be resorted to or the terreplein at the gorge lowered.

91.—To defilade a work from two or more heights, the plane must be tangent to the two heights to which angles of elevation are the greatest. As three points fix a plane, it follows that the tangent plane would usually contain but a single point of the string at the gorge, hence, the problem is solved by reversing the string triangle—i. e., fixing the apex at the gorge 3.5 ft. above the ground, and the two extremities of the base within the the proposed work and far enough apart to allow the two heights to be seen between them. An assistant at each of the forward stakes adjusts the string as directed. (Fig. 3.) The problem is then completed as in the previous case.

92.—It is sometimes advisable, when a single plane of defilade gives too great a command, to use two planes; the portion of the interior of the work on the side toward H (Fig. 4) being defiladed from it, and that on the other side from the height H'. This method exposes the faces and flanks to reverse fire and renders traverses (parados) necessary.

93.—The height of a traverse (which should be such that a shot grazing it will pass 2 ft. above the parapet it is to cover) is found as follows:

Assume that the traverse is to be on the capital of a lunette.

The problem of direct defilade with two planes having been solved, and the poles at angles of the works having been sawed off to indicate the proper height of interior crest in order to defilade the work as far as the capital, the height of traverse to protect a flank from reverse fire is found thus: Measure down from the tops of the poles at the extremities of the flank any convenient distance, as 3 feet,* mark the points and connect them by a string. This string and the opposite height determine a plane which will cut rods held vertically on the capital, at a distance of 5 feet below the required top of the traverse (2 ft. plus the distance measured down on the poles). Proceed in a similar manner, using the other hill and its opposite flank. The greater of the two results fixes the height of that portion of the traverse. In the same manner, its height to protect the faces from reverse fire may be found. By reference to Fig. 4 this will be readily understood.

This method, while not absolutely accurate, will give results near enough for all practical purposes, with the error on the side of safety. Traverses or Parados are the usual protection against reverse and enfilade fire, and, although sometimes used to protect parts of a work from direct fire, this is usually attained either by raising the parapet, by lowering the terreplein, or by both these methods combined.

94.—Profiling. After the trace of the work has been decided upon, the problem of defilade solved, the poles at the angles cut off as indicated, and the cross section of the parapet decided upon, the next step is to erect profiles which shall correspond to this cross section. These profiles are, if practicable, to be made of strips or battens, 1 in. x 2 in. and placed at intervals of about 10 yds. along each face and flank, as well as at each angle.

For parapets not over 6 ft. in height, stakes may at once be driven into the ground and strips nailed to them, but for higher parapets it is more convenient to make the profile on the ground, merely driving short pickets in place of the long stakes in the first instance. When completed, the profile is up-ended and

---

* The idea being to have the string behind which the observer stands, when looking towards the height, at about the level of the eye.

PLATE 12.

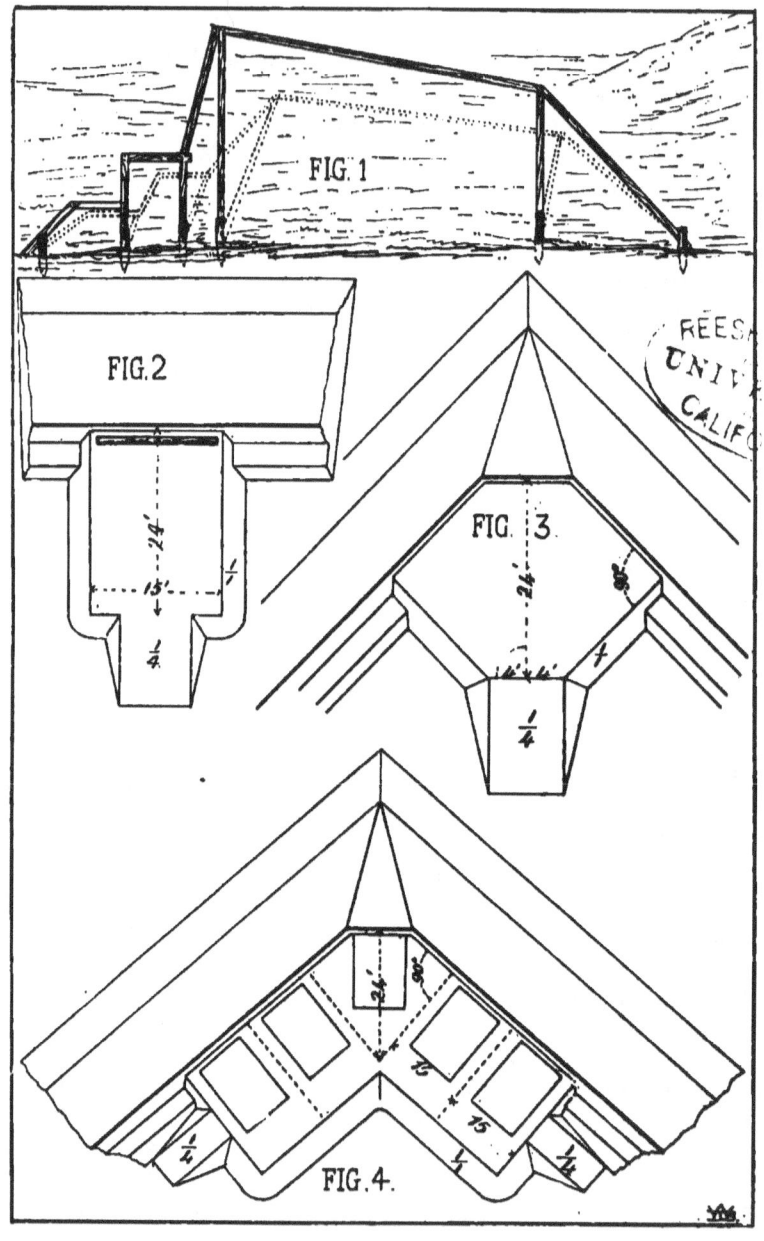

PLATE 13

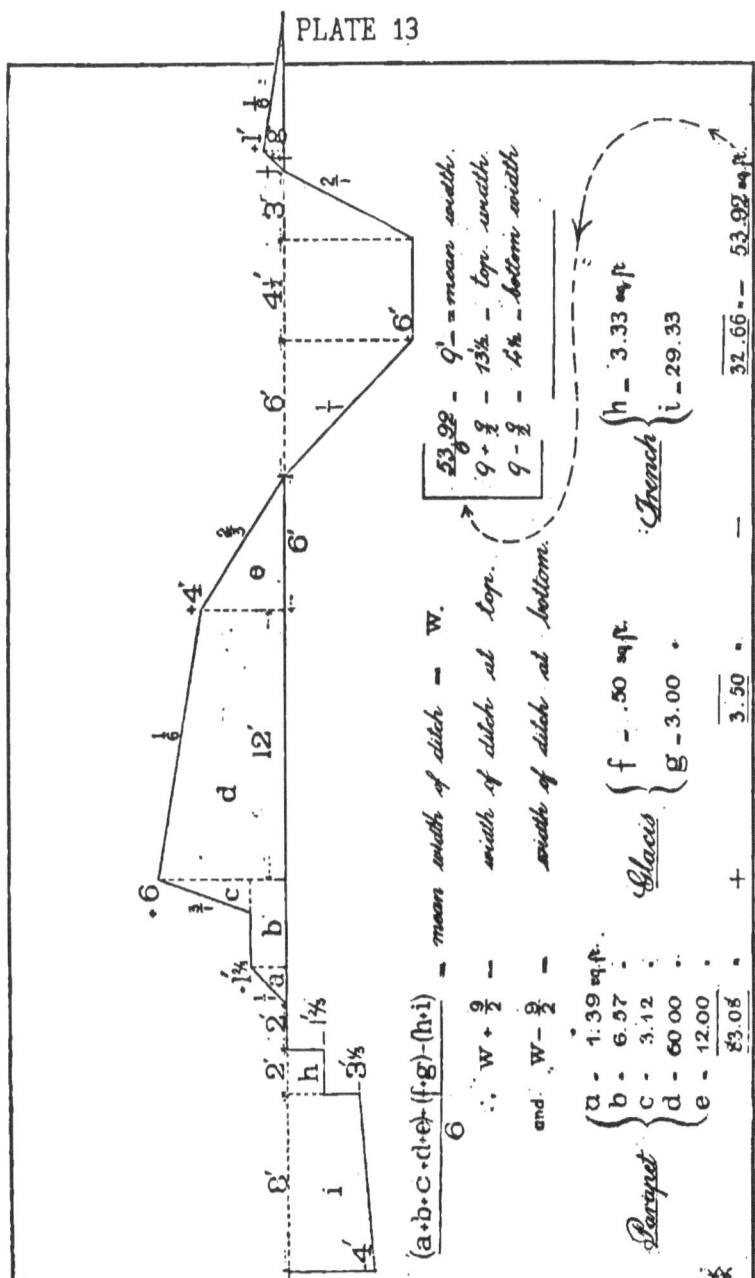

nailed to the pickets. (Pl. 12, Fig. 1.) If strips cannot be obtained, the entire profile, except the uprights, may be made of twine. The profiles at the angles of the works, known as oblique or angle profiles, will evidently differ from the others in length, while their height, on level ground, remains the same. The position of any point of the angle profile, as, for example, the exterior crest, is fixed by finding the intersection of the prolonged exterior crest lines of the face profiles. This result is accomplished by standing on the farther side of the second profile from the angle and lining in an assistant who holds a rod vertically at the angle, one end of the rod resting on the ground. After the profiles are in place, twine should be stretched between them to indicate the various crest lines. The outer edge of the battens marks the extent of the fill, except in the case of the interior slope, which is marked by the inner edge when the slope is to be revetted.

95.—Calculation of Dimensions of Earthworks. *The command* of the proposed work having been fixed by the requirements of defilade, and the *thickness* by the character of fire expected, *it becomes necessary to calculate the dimensions of the excavations, so that they will furnish enough, and no more, earth than is required.* The size of embankments and trench are, by the nature of the problem, fixed, as is the depth of ditch, hence the only variable is width of ditch, which is found as follows:

Assuming the relief to be constant and the profile, for example, to be as shown in Pl. 13, make a sectional sketch of the proposed work at any point except an angle. Calculate the sectional area of parapet, glacis, and trench, in sq. ft. and from the sum of the first two subtract the last: the remainder divided by the assumed depth of ditch, in feet, will give the mean width of ditch, from which, knowing the slope of escarp and counterscarp, the width at top and bottom can readily be found.

96.—Earth in embankment occupies, for a time, about one-twelfth more space than it did originally, but this increase is not usually taken into account in the computations for ascertaining the width of ditch. When the relief of a work is not constant, it is evident that, in order to get the proper amount of earth, either the depth or the width of ditch must vary. On account of the labor required in raising earth, the limit of depth is taken at 8 ft.; for a similar reason, the maximum height of parapet is taken at 12 ft. Whatever the depth of ditch assumed, it is always constant. The required width at any point is found by means of a section of the work, as already explained, a section near the extremities of each face determining the width of ditch for that entire face.

97. An excess of earth will occur at the salients and a deficiency at the re-entrants, although this may be partially obviated by making the shovelers throw toward the re-entrants.

98.—Drainage of the trench must be provided for at the time the work is constructed. If the fall is toward the gorge, an open drain will suffice, but, if in any other direction, a covered drain (Pl. 50) should be left to carry the water to the ditch.

Construction of Fieldworks. The details of construction and dimensions of earthworks will change with varying requirements and soil, but there are certain general principles that should be followed in all.

99.—As to profile : *The Normal* (Pl. 14, Fig. 2) fulfills the conditions as to simplicity, protection against field artillery (in most soils), command of the ground in front, and cover standing, in the trench. The trench is stepped and steps revetted to facilitate mounting the banquette, while the berm is omitted to deprive the assailants of a foothold. The command may be increased either with or without constant relief, the parapet thickened or reduced, and the trench made into a casemate without changing the type of this profile.

100.—As to garrison : For ordinary field works, the garrison is usually computed at 2 men per yard of interior crest; but for isolated works, this estimate should be increased by one-half. Embrasures and gun-banks each reduce the interior crest line available for troops, by 5 yards.

101.—As to laying out tasks : Cutting lines must be marked by tape or pick, computations made, and the exact size of the task for each relief determined in accordance with the rules given in Chapter VIII.

As an example of laying out tasks, assume that an earthwork with normal profile and constant command of 6 ft. is to be made on a level site.

Before work is commenced, the outer and the cutting lines of ditch and trench must be marked. As fatigue parties cannot be expected to excavate earth and at the same time preserve the proper slopes, the usual method followed is to dig vertically as in-

PLATE 14

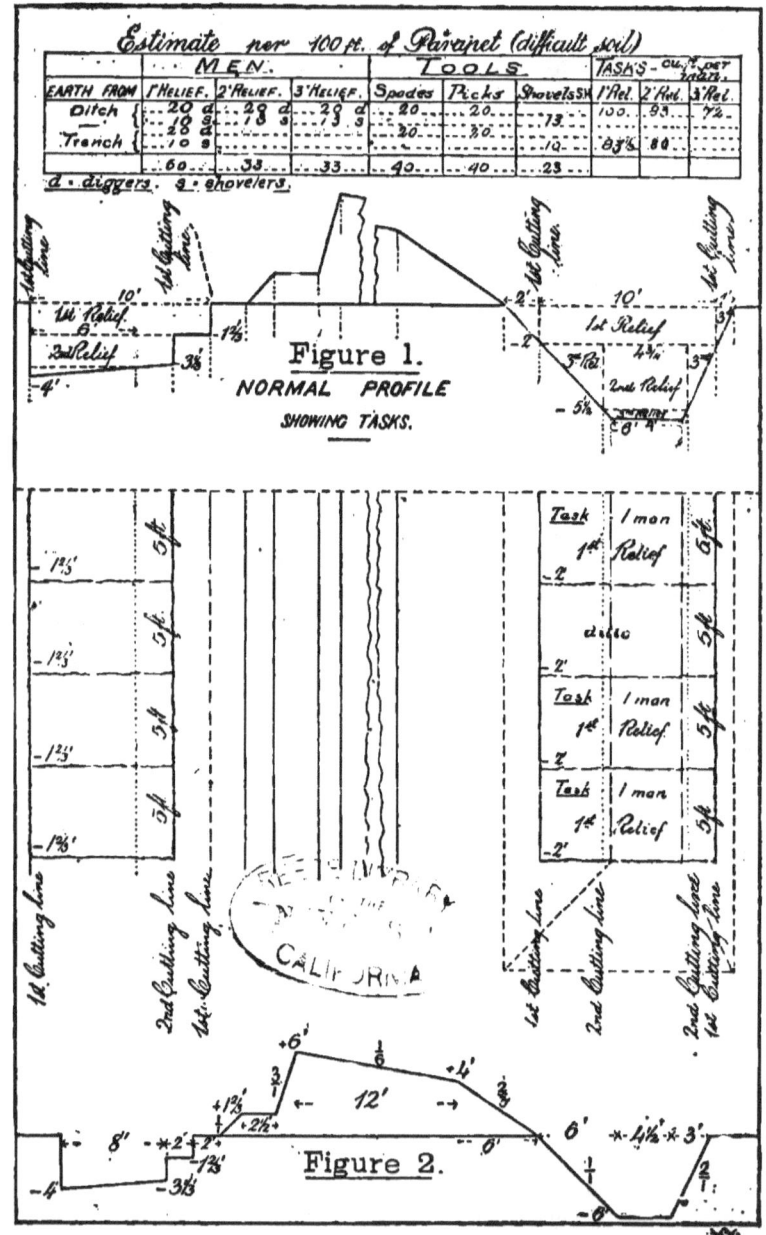

Figure 1. NORMAL PROFILE SHOWING TASKS.

Figure 2.

Pl. 14, Table, 2d relief, Trench, insert "20d" and "10s" and for total "33" read "63."

PLATE 15.

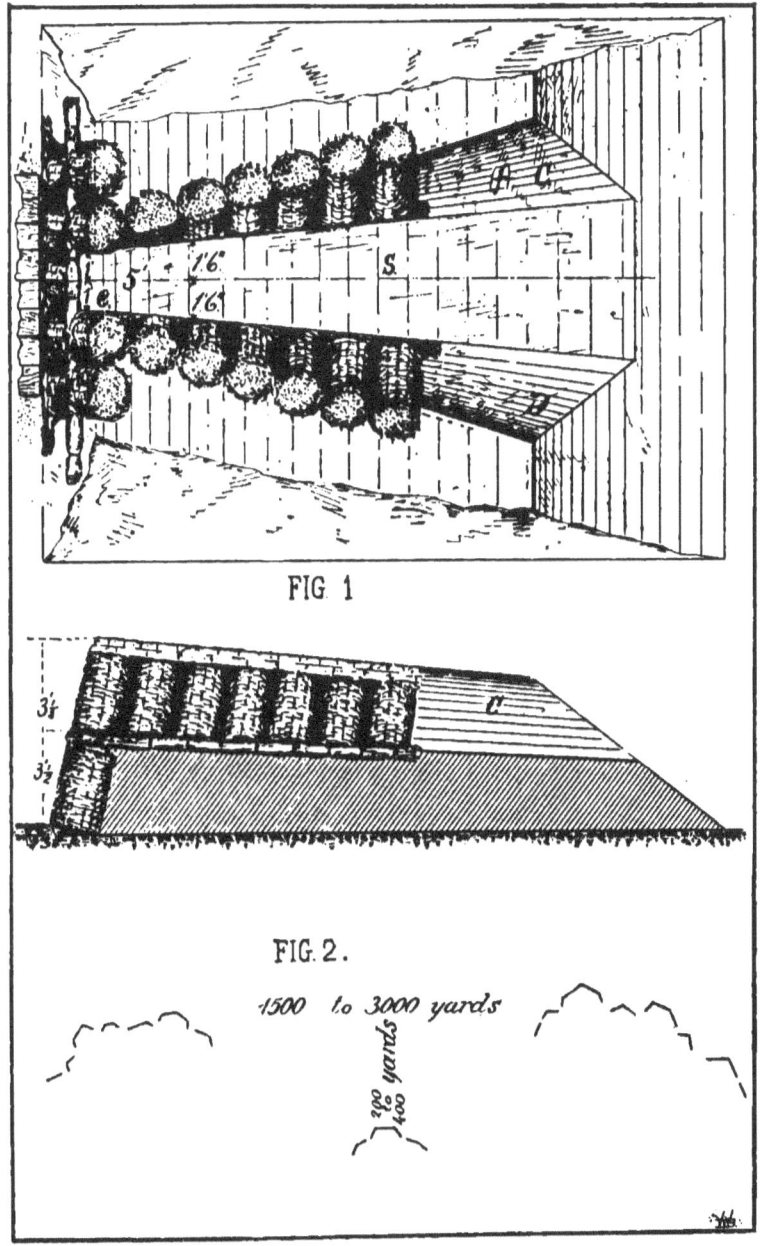

FIG. 1

FIG. 2.

FIELD WORKS. 67

dicated by the cutting lines and afterward form the slopes by cutting off the steps.

The cutting lines for the task of the 1st Relief would be made on the ground, as indicated in section and plan. (Pl. 14, Fig. 1). The 1st Relief having finished, cutting lines for task of 2d Relief would then be marked out, and finally, the 3d Relief would complete the slopes of ditch and parapet, and finish any work not completed by the other reliefs.

When not practicable to revet the banquette and trench steps, the risers may be sloped back at about six on one.

When necessary to throw earth more than 12 ft. horizontally, extra shovelers should be provided at the rate of 1 to each 2, or 2 to each 3 diggers, depending upon the soil and the distance it is to be thrown.

102.—Gun Banks, when made, are usually placed in the salients, for the reasons that the guns will have a greater field of fire and it is at this point that the earth of which they are made is in excess. Pl. 12, Fig. 2 shows a gun-bank on a straight line of parapet and Fig. 3 one at a salient. The top is horizontal and 3.5 ft. below the interior crest: this distance may vary, however, for different pieces. All slopes are one on one, except the ramp, or roadway leading up to the bank; this may be as steep as one on four, but a gentler slope is better. The width of ramp should be 8 ft. The level surface of the bank extends back 24 ft. from the parapet and a log or fascine is half sunken and picketed near the front, for a hurter. The width of bank for a single gun is 15 ft. At a salient (Pl. 12, Figs. 3 and 4) the angle is filled in by a straight revetment from 6 ft. to 15 ft. long and the superior slope reduced to correspond to lines joining its extremities with the exterior crest salient. This forms what is known as a "pan coupé."

103.—Embrasures for field guns would be used in positions where the fire is required to be in one direction only; for example, to sweep a road, bridge, or ford; or in the flank of a work to cover ground in front of an adjacent work.

Pl. 15, Fig. 1, shows the horizontal projection and the section of an embrasure. It is made at the same time as the parapet, by making the sole "s" parallel to the superior slope. The cheeks, "c, c," are vertical at the throat, "c," and have a slope of one on one at the other extremity; their height should never exceed 4 ft.

The usual method of forming an embrasure is to stretch a string along the line of fire; at the throat lay off 1 ft. on each side of it, and at a distance of 5 ft. from the throat lay off 1.5 ft. in a similar manner. Right lines joining the corresponding points so determined will mark the outer lines of the sole, which will splay one on ten. Each throat gabion is vertical, the extreme ones being inclined three on one; the slope of the intermediate ones is secured by alignment top and bottom on the extreme ones. Each gabion is anchored independently of the others, so that one may be torn out without seriously injuring the embrasure.

104.— By the **Merlon** is meant that portion of the parapet between two embrasures and above the soles. Embrasures should, as a rule, never be closer together than 15 ft., otherwise the merlon is too much weakened.

PLATE 16.

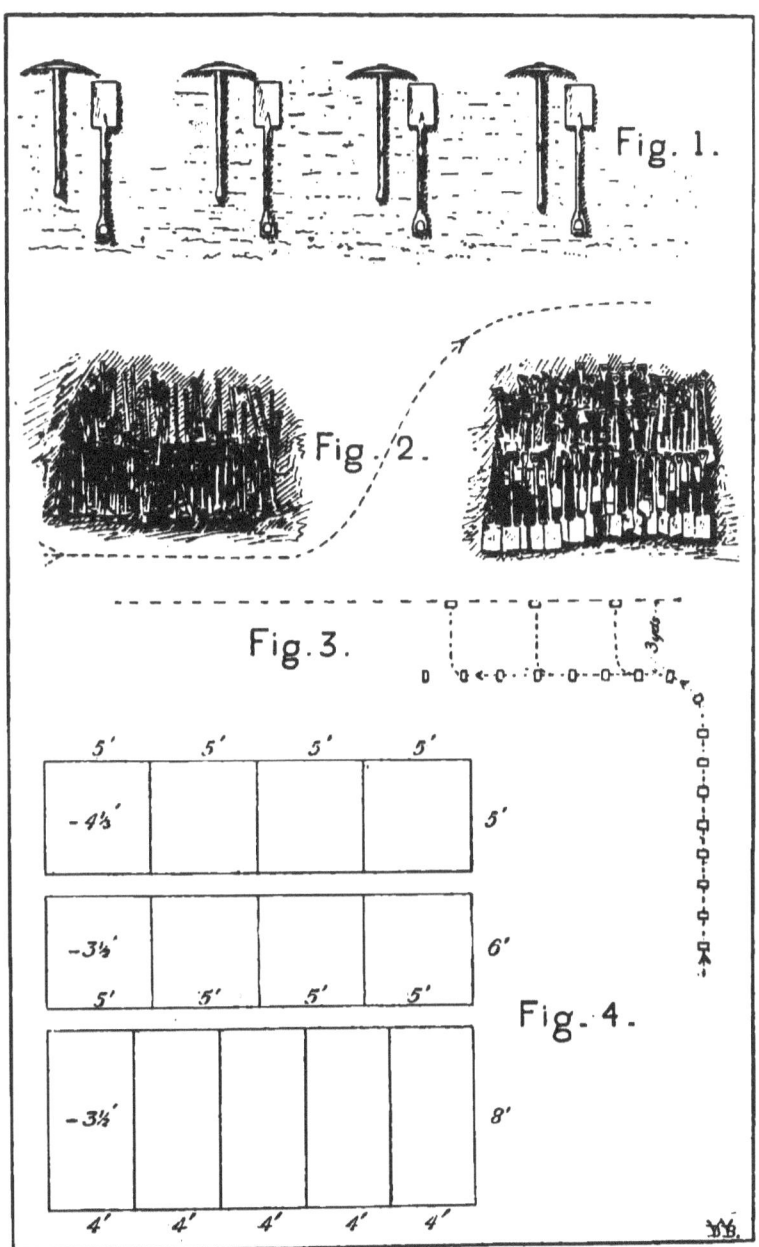

CHAPTER VIII.—Working Parties.

Occasion may arise, as, for example, at night, in the presence of an enemy or even with a large working party, when a well-established system of taking and handling tools, distributing and relieving working parties, etc., will be of paramount importance.

105.—Organization of Working Parties. The nature of the required work having been decided upon, the estimate of and the application for the requisite number of men and tools devolves upon the officer charged with its execution.

A working party of the requisite strength (which should include a reserve of 1-10th) should be furnished, as far as possible, from a complete organization, a company, a battalion or brigade, and not from detachments of different organizations.

106. Responsibility. The party should be divided into reliefs and the task each is to accomplish made plain before it begins work. The officers and non-commissioned officers of the working party are responsible for the amount of work done.

107.—Taking Tools. The first relief, having been formed in single rank with rifles slung across the back, is marched to the park where the tools have previously been laid out, either in rows (Pl. 16, Fig. 1) or in heaps. (Fig. 2.) The relief in the former case is advanced in line to the row and each takes a pick in the left and a spade in the right hand; in the latter case the party in column of files is marched between the piles, each in turn receiving a pick in the left and a spade in the right hand. The relief is then marched in column of fours or twos to the point where the work is to begin.

108.—Carrying Tools. In carrying picks and spades the handles are grasped near the iron, which is held vertically, the arms extended and the hands close to the side. In turning, the point of the pick should be lowered and the blade of the shovel raised, and when marching, either in line or in column, the handles should be splayed outward, in order to prevent interference. The necessity under certain circumstances of preserving silence makes the above precautions important at all times as a matter of training.

109.—Extending the Working Party. When the first relief approaches the designated point it is halted, then broken into column of files and direction changed, if necessary, so that the head of the column approaches in a direction parallel to and about 3 yds. in rear of the tape marking the front edge of proposed excavation. (Fig. 3.) When the leading file is opposite his place the command is given:

(1) *On right (or left) into line at two paces interval.* (2) **March.** (3) *Detachment.* (4) **Halt.**

The command "halt" is given when the leading file is 1 yd. in rear of the tape. While the line is forming, the correct positions are at once taken, as follows:

Each man on arriving at the line extends his arms horizontally, holding them thus until his own position and that of the man following him are established by touching hands. As soon as each man has his position he drives his pick into the ground on the left of his own task and lays his shovel on the ground, parallel to and at a distance in the rear of the tape equal to the width of his task from front to rear.

Rifles are then unslung, belts and canteens removed, and all having been placed on the ground three paces directly to the rear of task, butts of rifles toward the front, the men sit or lie down behind their shovels until the order *"Commence work."*

110.—Extension of 2d and 3d Reliefs. Each man of the 1st relief, after completing his task, scrapes his tools and lays them together in rear of the trench.

The task being completed, each man secures his accoutrements and rifle, and then, under direction of his officers, closes in to the left (or right), forming column of fours which is then marched back to camp.

As an incentive to rapid work, each relief should be allowed to return to camp on the completion of its task.

If the working party be large and the work of a complicated nature, each relief should arrive in successive detachments and

WORKING PARTIES.

their location on the work should have been previously designated, so that there need be no delay or confusion, even at night.

Work should not be commenced until the distribution of the entire relief is complete, since any change after work has begun tends to confusion, loss of tools, and delay.

111.—Tasks. An untrained workman can excavate in ordinary soil one cubic yard of earth per hour for four consecutive hours. As some men work slower than others, however, it is usual to estimate at 6 hours per man for the lifting of 4 cu. yds. of earth from a trench 3.5 to 5 ft. deep and throwing the same a horizontal distance of 10 ft.

When it is necessary to throw the earth more than 12 ft. horizontally, extra shovelers should be provided for re-handling it, in the proportion of 1 shoveler to every 2 diggers.

When exposed to the enemy's fire, a skirmish line is kept well to the front and the earth first excavated is thrown close to the edge of the ditch, forming a screen which is gradually thickened; under other circumstances the earth first excavated is thrown farthest.

Five feet, or two paces, is the usual distance apart for men to work, but they may be posted as close together as 4 ft., while using the heavy pick and shovel. As a precaution against injury to adjacent workers, the men should swing the pick in a direction perpendicular to the tape.

112.—Working parties may be extended at less or greater intervals by making the corresponding changes in the commands: when this is done, it will usually be necessary to verify intervals.

When necessary to complete a task in the shortest possible time or when the men available greatly exceed the number of tools, working parties should be formed in double rank, two men being assigned to each set of tools, which should be carried by the front rank man. When working in this manner with a double relief, the men, under direction of non-commissioned officers, should change off every 10 or 15 minutes.

Officers having general supervision of the work should not be changed at the same time the reliefs are.

The sizes of tasks based on the 4 yd. rule may be arranged as shown in diagram. (Fig. 4.) For arrangement of tasks in difficult soil see Pl. 14.

CHAPTER IX.—Revetting Materials and Revetments.

113. A Revetment is a facing used to hold up an embankment at a steeper slope than it would assume naturally.

114.—Revetting Materials. The revetments most commonly used in field engineering are made either of brushwood in the rough, fascines, gabions, hurdles, planks, timber, sods, sandbags, pisa, adobe, or of a combination of two or more of these.

115.—Brushwood, which is used in making the first four should be of willow, birch, ash, hickory, or hazel, and is most pliant when not in leaf: it may be of any size when used in the rough, but should not exceed an inch in butt diameter for gabions and hurdles and 2.5 in. for fascines and pickets.

The working party cuts and binds the brushwood in bundles of about 40 lbs. each, putting the large and small in separate piles with butts in the same direction. For convenience, the detail should be divided into three parts—one for cutting, one for sorting and binding, the other for carrying and, if necessary, loading. Tools required and time necessary are as in "Clearing the Ground" (Chap. V).

116.—Withes (Pl. 17, Fig. 1), which are used for binding and sewing, are made by twisting a pliant rod. The butt is held under the left foot and the twisting commenced at the small end, care being taken to avoid breaking or kinking the rod. The pliancy of the rod may be increased by heating it. In using the withes for binding, an eye is made at the small end, then the withe is passed round the bundle, the butt passed through the eye and twisted until a kink is formed, when the butt is thrust (buried) in the bundle.

117.—Fascines. A fascine is a bundle of rods tightly bound together. It has a length of 18 ft., a diameter of 9 in., and weighs about 140 lbs.

PLATE 17.

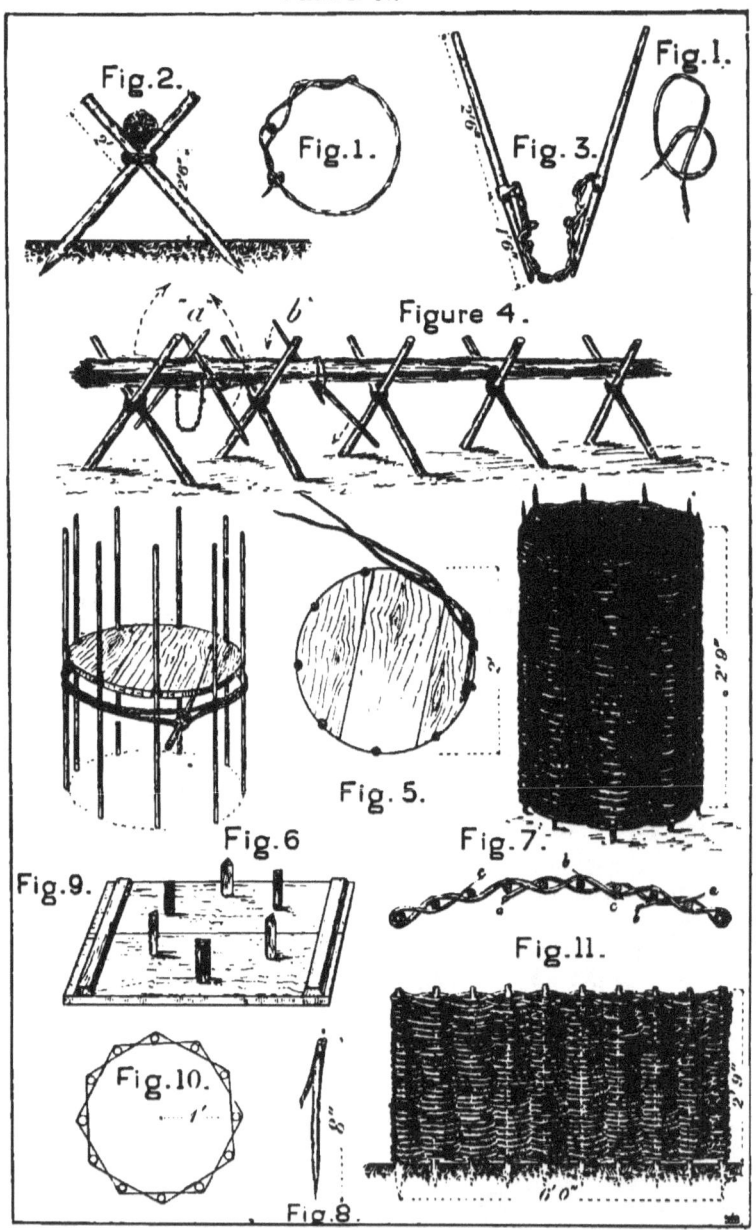

Fascine Rack.—The fascine is made in a cradle rack of five equidistant trestles (Figs. 2 and 4), the outer ones being 16 ft. apart, the crotches are each 2.5 ft. above the ground and aligned. The stakes for the trestles should be from 2.5 to 4 in. in diameter and from 5 to 6 ft. in length. Those for the outer trestles are first driven and securely bound together with wire or rope, then a line is stretched from crotch to crotch and the interior trestles made in a similar manner: the stakes should be driven firmly into the ground and each should have a length of 2 ft. above the crotch.

Fascine Choker.—For the purpose of gauging the circumference of the fascine and for cramping it in binding, the fascine choker (Fig. 3) is used. It consists of two stout bars or handspikes, 4 ft. long, to each of which is attached a collar 18 in. from the end, the collars being connected by a stout chain, to which are attached two gauge links 28 in. apart. The choker is used by a man on each side of the rack taking a bar of it and resting the short end on top of the fascine, chain being underneath (Fig. 4, "a"); then each passes his bar over to the other (the short ends passing around and under the fascine), and each bears down on the end of his lever. (Fig. 4, "b.")

Making the Fascine.—The trestles having been prepared, the fascine is made by laying brushwood, trimmed if practicable, in them, the pieces breaking joints and crooked ones being partly sawed or cut through. The rods should extend from 18 in. to 2 ft. beyond the extreme trestles and the bunch made of uniform size throughout. (Fig. 4.)

The choker should be used occasionally for testing the size and when of such dimensions throughout that the gauge rings meet, the fascine is bound. This should be done with wire or tarred rope, which is passed twice round the fascine and securely fastened, the bindings being 12 in number, the two outer ones 3 in. outside the extreme trestles and the others at intervals of about a foot and a half. This allows the fascine to be cut into lengths of 6 or of 9 ft. Five men require about an hour to make a fascine.

118.—Gabions. Gabions are open cylinders 2 ft. in exterior diameter by 2 ft. 9 in. in height, varying in weight from 35 to 50 lbs.: they are made of *brushwood, strap iron, iron bands* or *sheet iron* and from 9 to 14 pickets each. The interlaced brushwood in gabions is called the watling or web. Gabion pickets should be 3.5 ft. in length and from an inch to an inch and a half in diameter. The rods for the web should be from one-half to three-fourths of an inch in diameter, although smaller may be used. Wicker gabions are most easily made with the aid of a *gabion form*, which is a circular piece of board 21 in. in diameter, with equidistant notches on its circumference, the number of notches depending on the size of the brushwood and running from 9 to 14. (Fig. 5.)

The construction of the Wicker Gabion (Fig. 7) is as follows:

Watling. The gabion form is laid on level ground and a picket driven vertically in each notch, the thick and thin ends of the pickets alternating. The form is then slipped up the pickets about a foot and held firmly in place by means of a rope which is tied loosely round the pickets just below the form and then tightened by a rack stick (Fig. 6), the rope holding the pickets firmly in the notches. The rods for the web having been stripped of their leaves, the web is commenced by laying the butts of two rods in adjacent spaces between pickets, resting on the form. The rear rod, passing outside the second picket, is then bent inward, passing over the first rod, inside the third picket, and then out. (Fig. 5.) The other rod, which is now the rear one, is similarly treated and the watling continued by using the rods alternately. This method of watling is called pairing. On coming to the end of a rod a fresh one is laid alongside and woven with it for a short distance. The web is continued to within 3 in. of the ends of the pickets, care being taken to keep the pickets vertical and to make the web close by frequent use of the mallet.

Sewing. To prevent the web from coming off the pickets it is then sewed with wire, heavy twine, or withes, in four places, as follows: Take an end of a withe in each hand, the middle of it resting on the web, pass the ends of it through the web about 6 in. down the sides, one from without inward and the other from within outward; pull taut by bearing downward. Pass the ends through the web again 6 in. farther down and tighten as before. Proceed in the same manner a third time and then bury the ends of the withe in the web. The sewing should be at equal intervals and the two ends of the withe, when pushed through the web, should be separated by two or three of the watling rods; wire is much easier worked and more durable than withes. The partly completed gabion is now inverted, the form removed, and the watling continued as before, until the gabion has a height of 2 ft. 9 in., when it is completed by again sewing as before explained. The ends of the pickets that were driven into the ground are now trimmed to within 3 in. of the web and sharpened, the opposite ends sawed off to within an inch of the web, and a carrying picket driven through the sides of the gabion perpendicular to the axis and a few inches from it.

Three men should make a gabion in an hour.

119. Wicker Gabion Without the Gabion Form. Where the form is not at hand the wicker gabion is made by first describing on the ground a circle with a 10.5 in. radius and then driving the pickets at equidistant intervals on this line. The watling is commenced at the ground and run up to the full height, care being taken by frequent gauging to keep the dimensions accurate. It will be necessary for one man to devote his entire attention to keeping the pickets in position, while a second makes the web, and a third prepares the rods. Three men should make a gabion, without the form, in an hour and a half to two hours. Instead of sewing, the gabion may be finished by driving four forked pickets (Fig. 8) in the web alongside of the gabion pickets.

120. The Hoop or Strap Iron Gabion. This is more durable and more quickly made than the wicker gabion, but is heavy, weighing 55 lbs., and liable to splinter dangerously. The *form* for this gabion is used solely for gauging and shaping the bands.

To make the hoops, describe on a wooden platform a circle with a 1 ft. radius and divide it into 6 equal parts. Make auger holes at points of division and insert in them wooden pins about 5 in. long and triangular in cross section, the bases of the triangles being on the interior of the circle. (Fig. 9.) Wrap the strap iron once tightly round the pins, thus forming an hexagonal hoop. Mark the point where the hoop is to be joined, then remove, punch, and rivet it. As the iron is usually 1 in. wide the completed gabion will require 33 of these hoops.

To make the gabion, place a hoop on the ground and another on it in the positions shown. (Fig. 10.) Drive a picket vertically in each of the triangular spaces, then place the remaining hoops alternately over the first and second. Drive nails in four of the pickets outside the extreme hoops to keep the gabion intact.

121. The Sheet Iron Gabion. This gabion is made of a piece of sheet iron 2 ft. 9 in. x 6 ft. 4 in., riveted or wired together along its shorter edges.

122.—Hurdles. The hurdle is a brushwood mat 2 ft. 9 in. wide by 6 ft. long, the length corresponding very nearly to the circumference of the gabion. An even number of pickets, usually 10, is used in making it, the extreme pickets being somewhat heavier than the interior ones. (Fig. 11.)

Construction of Hurdles. Describe on the ground an arc with an 8 ft. radius, measure off 6 ft. of this arc and drive 10 gabion pickets along it at intervals of 8 in. (Fig. 11.) Commence the watling in the center space on the ground by randing, *i. e.* working with a single rod alternately inside and outside of the pickets; on reaching the end picket the rod should be twisted as a withe, so as to avoid breaking it, and then returned toward the center in the same manner as at first. When approaching the end of a rod another should be laid alongside of and randed with it for a distance of two or three pickets. Pairing, as in gabions, should be resorted to in finishing the top and bottom of the web and the hurdle should then be sewed as described for the gabion. When the rods used in watling are very small the process of slewing should be resorted to: this is the same as randing with the exception that 2 or 3 rods are laid alongside each other instead of using them singly. Slewing makes weaker work than randing. Three men should make a hurdle in two hours; two work at the web and the third prepares the rods. The completed hurdle weighs about 50 lbs. The hurdle is made on a curve and afterward flattened as much as possible, because it is found that by so doing it is less liable to warp than if made flat. It should be placed in a road or revetment with the concave side toward the earth.

123.—The Continuous Hurdle is usually preferred for revetting purposes to single ones joined. It differs from the latter in that the pickets are driven at once, at intervals of 12 to 18 in. according to their thickness, in the position the revetment is to occupy, but at a slightly gentler slope so as to allow for straightening when the earth is tamped. It is constructed by randing or slewing, two men being assigned a task of 10 or 12 ft. in length, which they should finish to a height of 4 ft. and anchor, in from one-half to three-quarters of an hour.

PLATE 18.

Revetments.

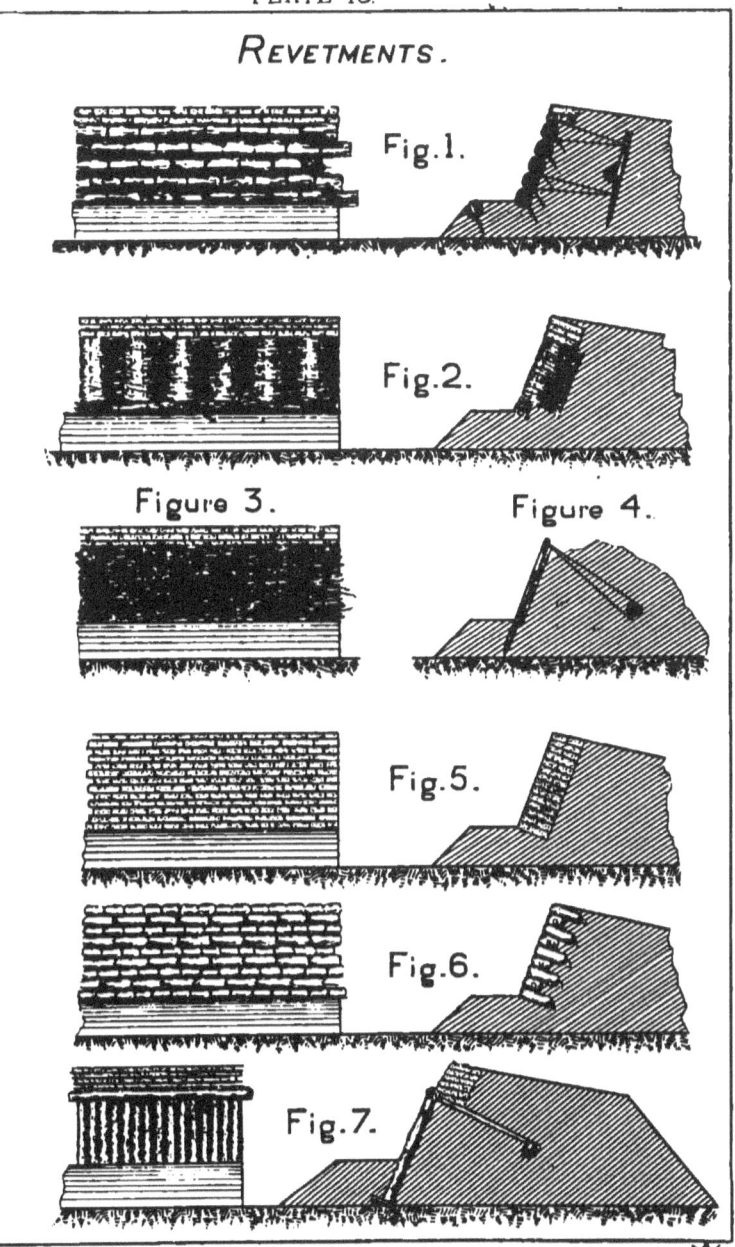

124. Planks when used for revetting should be placed edgewise and held in position by stout stakes which should be anchored. They make a neat, durable and quickly made revetment.

125.—Round timber from 3 to 8 in. in diameter may be used in the same manner as planks but the revetment is more difficult of construction and is not so durable.

126. Sod for revetting purposes is cut of a uniform size—18 in. long, 9 in. wide, and 4 in. thick. They should be laid in alternate rows of headers and stretchers, grass down, breaking joints, and perpendicular to the slope. The top layer should be all headers and have the grass up; alternate rows should be pinned securely, using split pickets, if possible, as with them there is less liability of splitting the sod than when round ones are used. Two men should lay from 70 to 100 sods per hour, depending upon whether or not pickets are used.

127.—Sand-bags are made of coarse canvas or bagging material and, when empty, measure 2 ft. 8 in. by 1 ft. 4 in. When filled they are supposed to contain 1 cubic ft. of earth; it is found in practice, however, that a cubic yard will fill from 48 to 50, making their average size 1 ft. 6 in. long, 10 in. wide, and 6 in. thick. Each bag has eyelet holes near the mouth through which a stout cord passes, to expedite tying, when filled.

For filling sand-bags the working party is divided into squads of 6: 2 with shovels, 1 with a pick, 1 to hold the bag, and 2 to tie.

Each squad fills 150 bags per hour. This task may be considerably increased, however, in easy soil or with trained men, and the rapidity of the work more than doubled by having a double relief and keeping the men constantly changing.

128.—Revetments. Brushwood Revetment is made by driving pickets at intervals of about 12 in. along the foot of the proposed slope. The top of the pickets when driven should be as high as the proposed revetment, and the pickets should be anchored by wire to logs or stout stakes in the parapet. Loose brushwood is laid closely behind the stakes and earth tamped against it, the construction of the parapet going on at the same time.

Brushwood revetment is rapidly made in daylight but is neither durable nor sightly.

129.—The Fascine Revetment. (Pl. 18, Fig. 1.) This is

made by laying the fascines in single rows of stretchers, breaking joints, each fascine being pinned to the parapet by 5 or 6 pickets, and every second or third row securely anchored.

Six-foot fascines should be used occasionally as headers. The bottom fascine is sunk about one-third of its diameter by excavating a shallow trench. The construction of parapet and revetment proceed simultaneously. Slope should not be greater than four on one. The defects of this revetment are the weight of the facines, the large quantity of brushwood required, and the fact that the fascines are held in place by anchors and pickets in the earth which they support.

130. **The Gabion Revetment.** (Fig. 2.) This is made by first sinking a row of fascines about 3 in. at the foot of the slope, so as to give an inclination of four on one to the gabions resting partially on them. Earth is tamped behind and in the gabions, and sod or sand-bags placed on top. Where greater height is required two rows of gabions may be used with two facines, well picketed, between them.

Gabions make one of the strongest and most durable revetments, their own weight when filled being usually sufficient to retain the embankment.

131. **Hurdles.** These make a poor revetment unless the method is followed of constructing a "*continuous hurdle*" at the same time with the parapet. To do this, the pickets are driven along the foot of the slope at an inclination of about three on one, when the final slope is to be four on one. The watling is made continuous by randing or slewing, each two men having four paces of hurdle as a task, and taking care to work in their rods with those of adjacent sections. (Fig. 3.)

132. **Plank or Timber Revetment.** (Fig. 4.) This is made by driving heavy stakes into the ground at the proper angle, placing the planks or timbers behind them, then filling in and tamping firmly. The stakes must be anchored. This revetment is neat and durable.

133. **Sod Revetment.** (Fig. 5.) This is made by laying the sod in alternate layers of headers and stretchers, grassy side down, breaking joints and perpendicular to the face of the revetment. Each sod should be well settled before another is placed on it and the top layer should be headers with grass up. It is well to pin alternate rows by means of split pickets, three

fourths of an inch in diameter and 9 in. long. This revetment is made of uniform thickness throughout by using double rows of stretchers. If the grass is long it should be mowed. If the sod is very wet when laid the revetment will crack in drying. Two men well supplied with sod should lay two paces of revetment, four and one-third feet high, in an hour.

This revetment has the advantage of not splintering like gabions, fascines and boards, but should not be used when other material is obtainable because ordinarily it will not stand long at a steep slope (three on one being about the limit), cannot be used when very dry or frozen, and requires great care to build properly.

134.— Sand-bag Revetment. (Fig. 6.) This is made by laying alternate rows of headers and stretchers, breaking joints, and perpendicular to slope, seams of stretchers and chokes of headers being put in the embankment. Men working in pairs lay the bags, settling them firmly in place with a mallet or spade. This revetment is not very durable but the bags are easily transported, may be used with any soil, and are invaluable in making hasty repairs and loop-holes.

135.—A very durable revetment, (Fig. 7,) much used in the defences of Washington 1861-5, was made of posts (oak, chestnut, or cedar) cut in lengths of 5.5 ft. and placed side by side, at a slope of six on one. The footing was a 2 in. plank laid in a trench excavated for the purpose. The tops of the posts were sawed off 16 in. below the interior crest and capped by a half-round timber, all being securely anchored in the parapet. Crowning was completed to the requisite height with sod.

All revetments that are liable to splinter should be crowned to a height of at least 8 in. with sods, sand-bags or earth.

136.—Pisa Revetment is made of earth and clay, to which has been added enough water to reduce the mixture to a working consistency. A trench 6 in. deep and 18 in. wide, is first dug, its nearest edge marking the foot of the revetment. Pickets, of sufficient length to reach the top of the proposed revetment, are firmly driven, at the proper angle, about 2 in. from the near edge of the trench, at intervals of about a yard, and then anchored. Boards placed horizontally are now laid against the pickets on the trench side. The trench is then filled with the mixture, tamped, and more added, other boards being placed on top of the first, as required, and the mixture forced closely against them The construction of the parapet goes on at the same time with the revetment. When completed, the pickets and boards are removed, This revetment is neat and durable but cannot be rapidly made.

137.—The Adobe is a sun-dried brick, about 18 in. x 9 in. x 4.5 in. and when carefully laid with the same bond as given for sod or sand-bag, forms a neat and very durable revetment, exceeding in the latter respect any of the other varieties mentioned.

The following table shows amount of various materials required for 100 running ft. of 4 ft. 4 in. high revetment.

Kind of Revetment	Fascines	Gabions	Sod	Sand-bags	Pickets
Facines	30		267		150
Gabion	6	50	400		
Sod			1867		1000
Sand-bag				867	

PLATE 19.

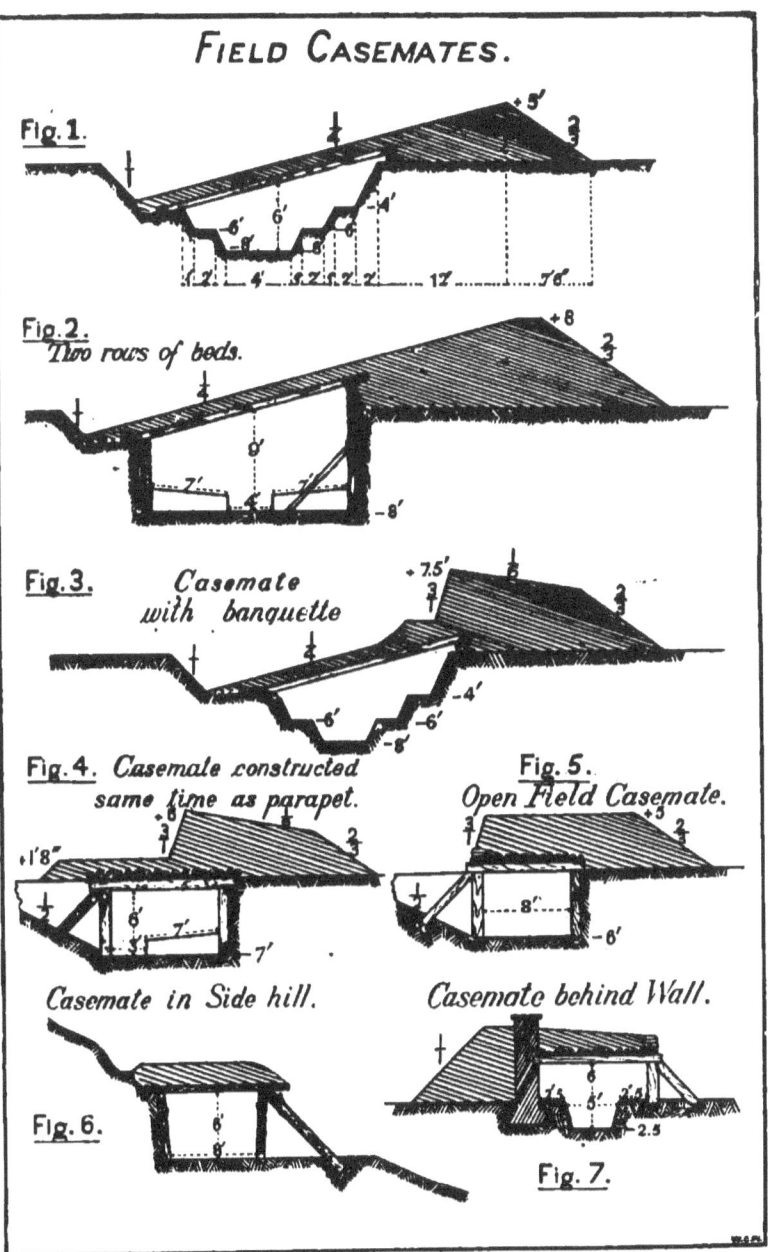

CHAPTER X. Field Casemates and Magazines.

138.—In all field works, protection against both weather and hostile fire must be provided for the garrison.

These shelters are constructed by building a chamber of wood sufficiently strong to bear the necessary earth covering, and by protecting this in front by an embankment thick enough to withstand direct artillery fire.

Two general forms are used:—

(1) Those which, after providing complete protection from direct fire, have their roofs sloped to the rear at an angle greater than the angle of descent of the enemy's projectiles, generally about one on four; and,

(2) Those which have horizontal roofs, the earth covering being so high and massive as to protect against artillery fire by its thickness alone.

The first class is preferable, the work of construction being very much less than in the second class, as the embankment is not so high and the earth on the roof does not require to be thicker than 16 in., as it has to resist only the dropping fire of small arms and the fragments from bursting shrapnel. Moreover, it gives much easier drainage to the ditch in rear.

139.—The construction of the timber part of the casemate is practically the same in both cases. The vertical timbers being rough tree trunks, about 1 ft. in diameter, placed at intervals of 3 or 4 ft., and strutted when necessary. The roof timbers in simple casemates being not less than 8 in. in diameter and the interstices filled with small poles or brush. In case the protection has to be proof against vertical fire of mortars, the earth mask on the roof must be 6 ft. in thickness and a correspondingly stronger timber construction must be provided: these are shown in Pl. 19, Figs. 1 to 7.

In calculating floor space, each man should have from 9 to 18 sq. ft.; the former when crowded, the latter when not.

140.—Magazines are of two kinds: *First*, those intended to hold the temporary supply for guns or troops when in action; and, *Second*, those intended for the purpose of storing ammunition in large quantities.

The first variety consists of recesses in the interior slope of the epaulement—barrels or gabions are excellent and when not obtainable may be replaced by empty ammunition boxes placed in holes excavated for their reception.

Magazines of the second class are used only in works of great defensive value and then only when ample time is available. They are made in the same general manner as the casemates heretofore described, except that great care must be taken to render the structure as dry as possible and to secure good ventilation.

141.—The general plan of execution of these works is as follows:—

(1) Magazine shown in Pl. 20, Fig. 1.

The mask in front should be 20 ft. thick. The roof consists of a row of timbers or logs 8 in. in diameter, overlaid with steel rails, and then covered with a paulin, well tarred if possible. On this is placed 16 to 18 in. of earth. The ends are made of logs, 12 in. in diameter, planted in a double row, breaking joints. The entrance is at either one or both ends according to circumstances. The doors, 2 ft. 6 in. in width, are made of planks crossed, and are hung next to the front wall of trench, opening into a passage formed by a row of upright logs parallel to those on the end of the magazine. At the end of the passage farthest from the first door a second one is hung, opening into the magazine. The vertical timbers in front and rear of trench support a revetment of planks or hurdles. The floor should be raised at least 6 in. from the bottom of the trench, to guard against dampness. Care should be taken to facilitate the draining of all water that falls on the roof, and that the trench itself is drained away from the ends of the magazine.

142.—Another form is as follows:—Determine the space needed for storage of ammunition. Then build the timber work as in the preceding, first excavating to a depth of 4 or 5 ft. over the entire site. There will be no ends to be closed by timbers.

PLATE 20.

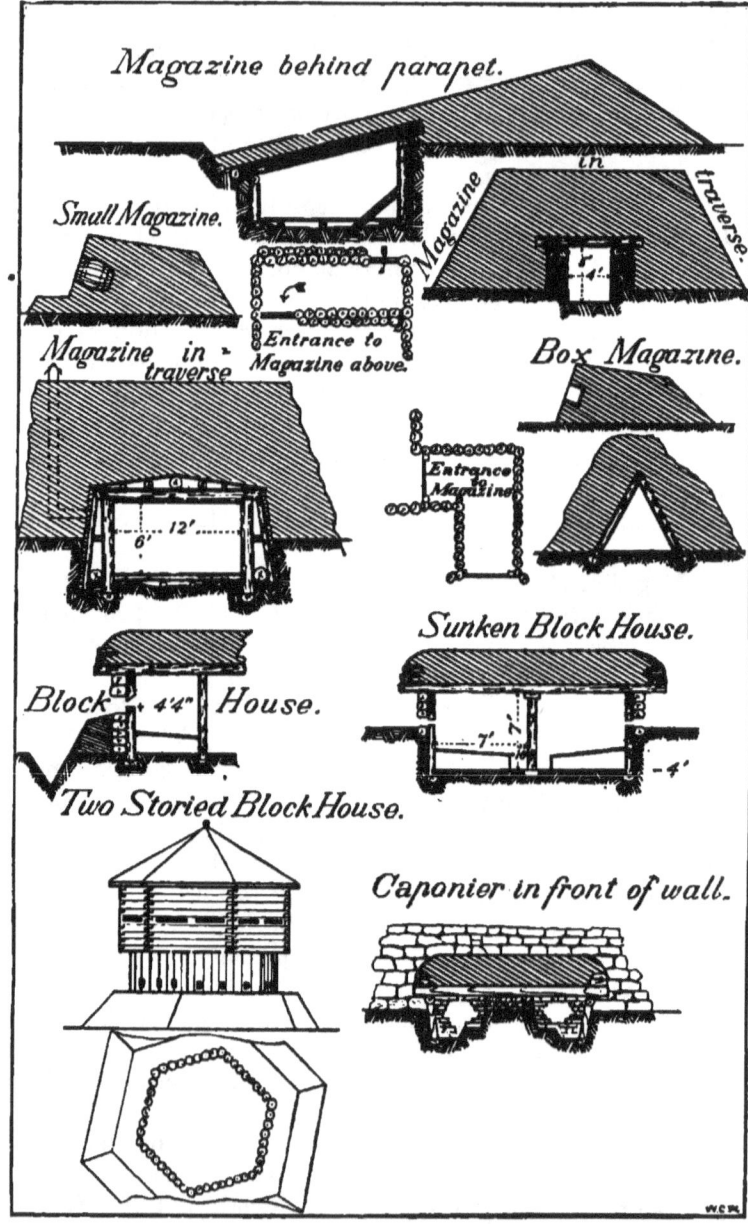

FIELD CASEMATES AND MAGAZINES. 93

The roof is made of timbers 12 in. in diameter, well supported by uprights of same size and long enough to give sufficient head room. The sides and ends should be revetted with plank, if possible, and the floor raised 6 in. above the earth. The center of the roof is raised a foot above the sides and surmounted by a layer of 6 in. of earth, well tamped; over this is laid a paulin and the earth mask is then placed over all to the thickness of 8 ft.; the covering mass in front should not be less than 20 ft. in thickness. Entrance is gained by means of a doorway opening into a passage which communicates through a return with the interior of the magazine. Doors made of crossed planks are hung as indicated in the plan. If time is available, and the planks at hand, an interior chamber should be formed, leaving an air space around the magazine proper; and inlets may be constructed, care being taken that they are not situated in exposed positions and that their course is such as to prevent the entrance of sparks. The roof should be rounded off so as to afford the easiest drainage. If the earth excavated is not sufficient to cover the roof, the necessary amount may be taken from a trench dug around the outside.

This form of magazine may with advantage be placed in a traverse.

143.—In case timber is not at hand, gabions and fascines may be used to build the magazine in the manner shown in Pl. 20.

144.—Block Houses are defensible shelters for infantry, although, under certain circumstances, they contain artillery.

They are generally used for the purpose of flanking defenses whose fire cannot reach into the ditch.

They are constructed either of upright timbers set in the ground close together, or horizontal timbers laid one upon the other; the timbers being in two rows, breaking joints in each case, or, if both methods are used, the outside row should be horizontal and the inner vertical. They should have at least 6 ft. of head room and should not be less than 9 ft. wide, as this allows one row of beds only. The roof should be of solid construction and covered with earth to the thickness of 2 ft. and should project 2 ft. over the wall to protect from dropping fire.

The walls should be masked with earth as high as possible and a ditch dug around the entire building. Loop holes are made at the height of 4 ft. 4 in. and are cut according to circumstances, as described in Chap. XIII. If necessary, block houses may be sunk in the ground, but a limit of 4 ft. in depth should be observed. The shape will conform to the necessities of the case.

145. In isolated positions they are advantageously made cruciform, thus presenting an opportunity for flanking each face of the house. When in wooded and mountainous countries, where artillery is not to be feared, these houses may be made with two stories, built so that the angles of the upper story project over the sides of the other, forming a machicoulis gallery, thus preventing the occupation by the enemy of the dead space in front of the straight walls.

146.—**Caponiers** are sunken block houses placed in the ditch of fortified places to prevent their occupation by the enemy: they are loop-holed about 18 in. from the ground, so as to have the most effective plane of fire. (Pl. 20.)

147.—**Tambours** are essentially block houses, having for their object the protection of angles, and the flanking of sides of buildings, and are especially useful in defending doors of buildings.

CHAPTER XI.—Field Works in Combination.

148.—Where several field works are used in conjunction, either as an intrenched position or in the investment of a fortress, city, or other important point, they constitute what is known as a *Line of Works*.

A **Line of Works** may be *continuous*, that is, forming, together with natural obstacles, an unbroken line, or, *with intervals*, by which it is understood that the works are distinct, either supporting each other or not, and the spaces between them not impassable by reason of natural obstacles.

149.—**Lines with intervals** have the following advantages over *continuous lines*, viz:—

(a) They involve less labor.
(b) The garrison of the defenders may be smaller.
(c) They allow greater freedom of movement for counter attacks.

The general principle to be followed in their construction consists in forming a line of fortified points or pivots. These points or pivots detain the enemy's advance, since he would hardly pass them and expose his flanks and rear, while a continued unsuccessful attack on the strongly fortified pivots would open the way for a counter attack by the defenders.

When, however, the defense is intended to be solely passive, which would be the case while awaiting reinforcements, or when the enemy greatly outnumbers the defenders, the intervals would be obstructed by felling trees or using any available obstacles, since counter attack is not contemplated.

In the use of lines with intervals, if the general defensive line is straight the works could be blunted lunettes with flanks traced so as to protect the front of adjacent works. If on a convex curve, the capitals should radiate from a common center, while on a curve concave toward the enemy, the capitals should converge and the front of each work might be a straight line.

When impracticable to construct the main works of a line with intervals, within supporting distance (600 yds. for infantry and 2000 yds. for artillery)* of each other, intermediate works retired from the main line, not more than half the interval, may be used.

In Pl. 21, Fig. 6. is shown such an arrangement, the pivots being single works while the artillery is retired from the main line and supported by infantry in shelter trenches.

Where the interval is as great as 1,500 yds. it is advisable to strengthen the pivots considerably, forming *groups*, the individual works of each group being so traced as to afford mutual defense. (Pl. 15, Fig. 2.) Each group in this latter arrangement forms a strongly fortified point of support and would usually have sufficient strength in itself to resist assault.

150.—Sometimes, when the defense of a line is of vital importance to the defenders, a *double line of works* is employed, the front line being shelter trenches or open field works of slight profile, the second line, not over 500 yds. in rear of the first, being field works of strong profile.

151.—Artillery should, as a rule, be placed outside of and somewhat retired from the works and protected by their own gun-pits or epaulements, for the reasons—

(1) That the works gain much in simplicity and rapidity of construction.

(2) That this disposition draws the enemy's artillery fire from the works and renders it more scattering.

(3) Greater mobility is given to the defender's artillery in case of advance or retreat.

(4) A better tactical position for this arm can often be secured than the one selected for infantry.

152.—As examples of continuous lines, Pl. 21, Fig. 1. is known as the redan trace with curtains. Fig. 2. is a modification of Fig. 1., the redans being blunted. Fig. 3. is the tenaille trace. Fig. 4. is a tenaille and redan trace. The crémaillere trace (Fig. 5) has long faces and short flanks.

With respect to the continuous lines above mentioned, the

---
* Continuous dangerous space for new Infantry rifle (Krag-Jorgenson) is 610 yards.

PLATE 21.

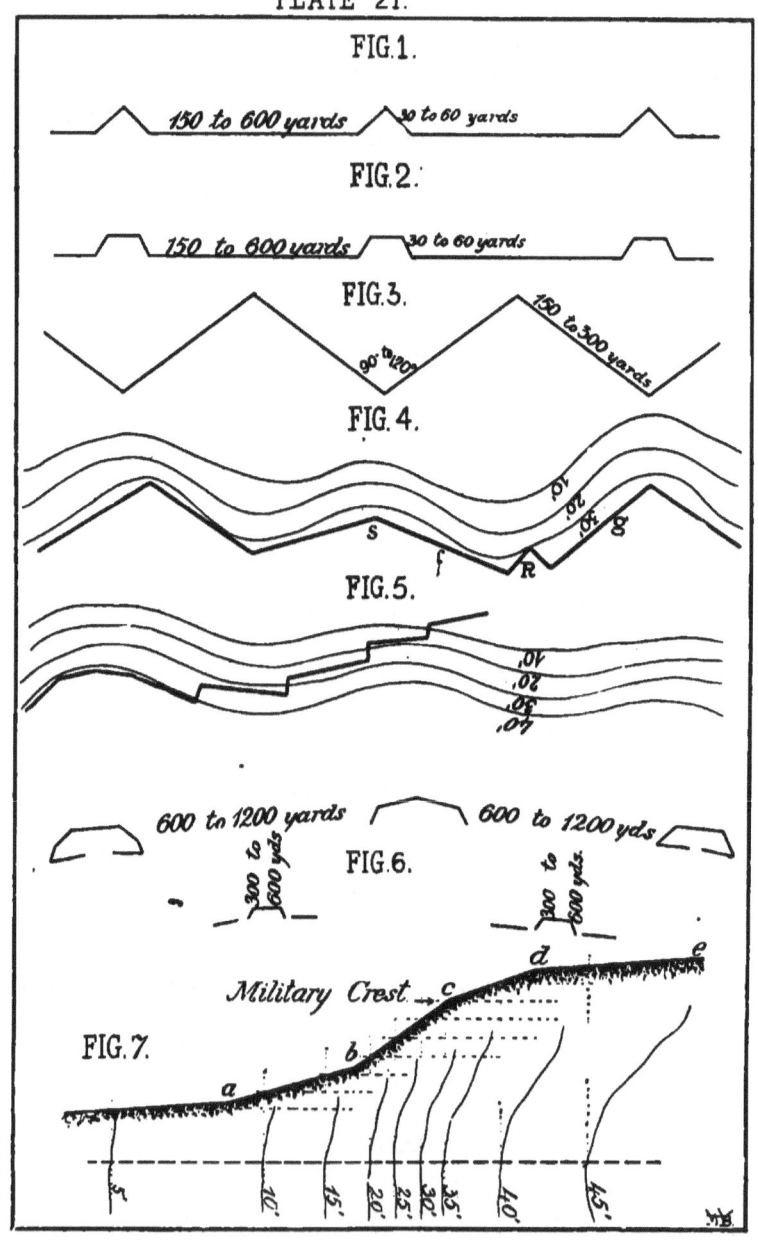

preference on a level site would usually be given to the trace shown in Figs. 1 and 2, the artillery being placed in the most favorable position along the curtains, with machine guns in the most important redans.

The tenaille and the tenaille and redan trace (Figs. 3 and 4) are objectionable, in that they involve more labor, cannot bring as direct a fire to the front, and the faces are liable to enfilade when the salient angles approach 90°, while, on the other hand, if a salient, as "S," Fig. 4, approaches 120°, mutual defense of the faces, "f" and "g," would be lacking, thus making the redan, "R," necessary. This trace may, however, be rendered unavoidable by the conformation of the ground.

The crémaillere trace finds special application in a position such as is indicated in Fig. 5, viz., joining two points, one at the top and the other at the bottom of a slope, the short flanks affording but limited opportunity for enfilade fire.

153.—The strength of a defensive position lies in a great measure in the proper utilization of the accidents of the ground; thus, the traces that have been mentioned may have to undergo considerable modification to be appropriate to the varieties of terrain constantly met. It is evident that, in a broken or hilly country, one by preference would occupy the heights. These, from a tactical point of view, possess the advantage of overlooking the low ground in front, besides the great advantage of concealing from the enemy the movements of our own troops in rear; but, since all else must be subordinated to fire effect, it is evident that such a line on the heights should be selected that the defenders may completely cover the ground over which the enemy must approach. This naturally leads to the inquiry as to how that line may be determined.

Heights, great or small, usually present the profile shown in in Fig. 7, that is to say, they have a steepest slope, "b c," which is joined to the crest and to the valley below by the two gentler slopes, "a b" and "c d." In order, then, to beat the zone "b c" it is necessary to occupy the crest "c" or some point below it on this slope. To distinguish this crest from others, it will be called the *military crest*.

With the inclination of this steepest slope greater than one on four, it is unusual to construct anything but shelter trenches along the military crest, the artillery being retired sufficiently and placed in such positions as to command a good view of the rest of the field. With gentler slopes, however, the artillery may be placed at intervals along the military crest, the intermediate spaces being held by infantry in shelter trenches.

A better disposition than this, where the ground permits of it, is to place the infantry trenches part way down the slope in front of the military crest, the artillery occupying a position in rear of and close to the crest, so that little more than the muzzles of the pieces are visible. In this case care must be taken that the infantry trenches do not mask the fire of the artillery.

In choosing a defensive position the ground should be viewed from the highest point in the vicinity and by frequent practice the eye so trained that the military crest is at once apparent and the slopes instinctively classified with respect to their use by the different arms.

Finally, the distance to a number of visible permanent points in front of the works should be determined and recorded, so that there may be no necessity for range finding during the enemy's advance.

PLATE 22.

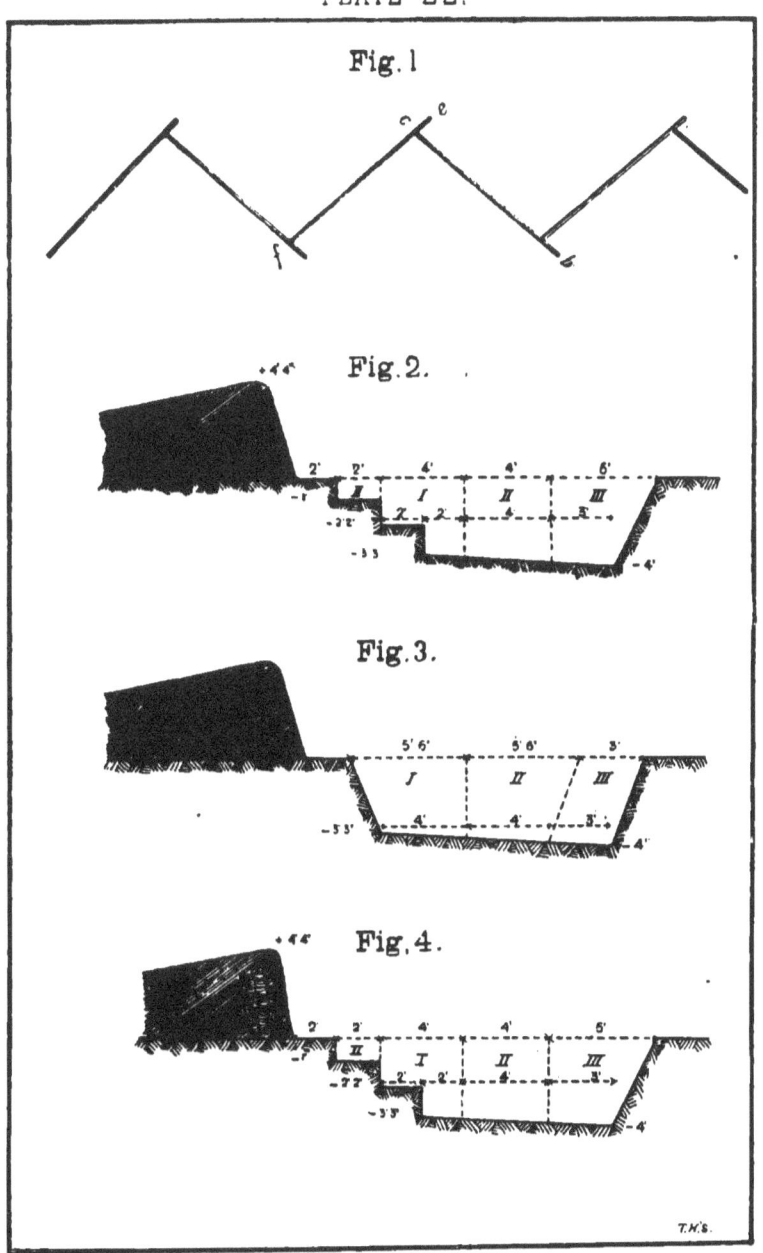

CHAPTER XII.—Siege Works.

154.—When it becomes necessary to besiege a place, it may be approached by common trench work or by some form of sapping. As the common trench and the flying sap are the work of infantry, they alone will be referred to.

155.—The method of providing the working party with tools, laying out the work, and extending the working party is described in Chap. VIII. It is to be noted, however, in work of this character, *that when extending along a zig-zag, upon reaching the angle the order of forming up must be reversed.* Thus, if the column from b to c (Pl. 22, Fig. 1) were forming on the left, upon reaching e f, it would form on the right.

156.—Common Trench Work. This may be used as a parallel, an oblique approach, or communication. The work done by reliefs in constructing a parallel is shown in Fig. 2. In this case, as musketry fire must be provided for, the second relief cuts out the top step. Should it be necessary to revet the bottom step, fascines for this purpose may be carried by the second and third reliefs. This may be and usually is omitted until the parallel is completed. Fig. 3 shows the common trench used as an oblique approach or communication. Should the trench be found wide enough the task of the third relief may be omitted.

157.—The Flying Sap (Fig. 4) is similar to common trench work, except that in the former case the embankment is revetted with gabions.

In taking tools, each man of the first relief is, in addition to his pick and shovel, provided with two gabions. In laying out the tools, a shovel should be fastened in one gabion by being placed between two of the gabion pickets, handle of the shovel inside. A pick should be secured in the other gabion by having its point pushed under the pairing rods, handle inside. The gabion with shovel is taken in the right hand, the one with pick in the left: both gabions being carried by carrying pickets.

158.—The extension in the flying sap is made from single rank on the right or left, and differs from the extension in common trench work in that the interval, in the former case, is the width of two gabions. Each man, on coming into line, places his gabions so that they touch each other along the inner edge of the tape, takes out his tools and lays them down, as explained for working parties in Chap. VIII., and waits for the command, "Commence work." In commencing work, the gabions should first be filled, hence the position of each pair of gabions should be rectified before this command is given.

159. —Each branch of a zig-zag should receive such direction as not to expose it to enfilade fire from any point of the defenses. Its prolongation should, therefore, fall outside of the most advanced salient of the collateral works. Ordinarily it would not be longer than 100 yds. A parallel should be stepped in order to facilitate an advance from it.

160. A portion of the parallels and approaches used in the capture of Fort Wagner, Morris Island, S. C., September 7th, 1863, is shown in Pl. 23.

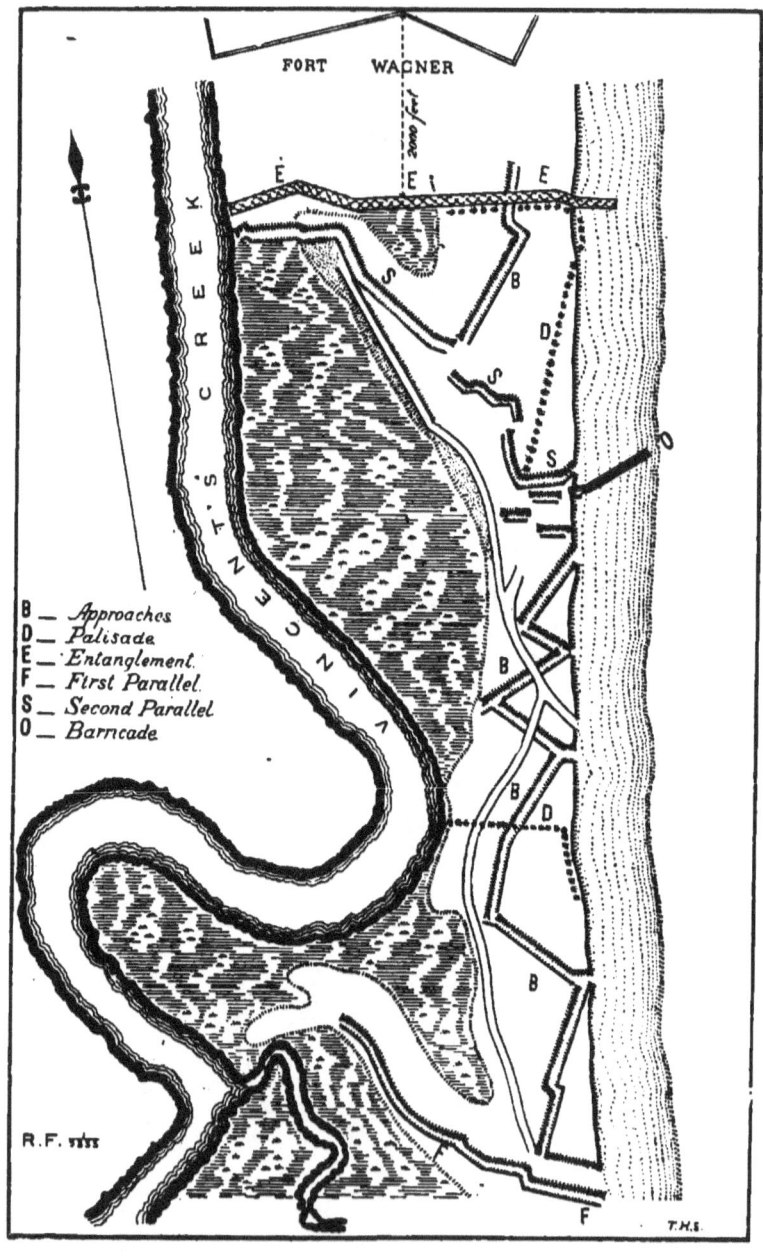

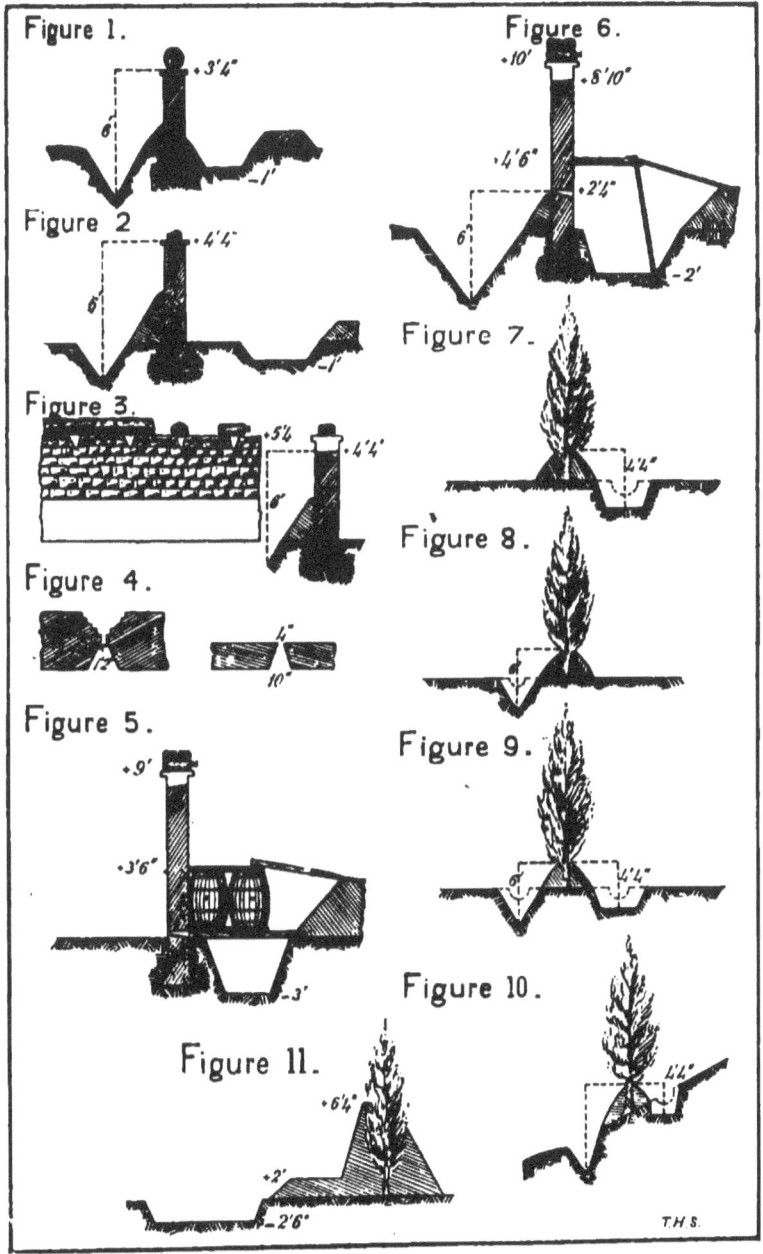
PLATE 24.

CHAPTER XIII.—Defense of Localities.

161.—Walls. Should the enemy close on them, walls must be so prepared that they will neither screen nor cover him, nor permit his firing from them. To prevent this, obstacles may be placed in front, or a ditch may be dug outside which will place him so far below the top of the wall or the bottom of the loop holes that he cannot fire over the one or through the other. A total height of 6 ft. will prevent this, or, in case of loop holes close to the ground, the maximum height should not be greater than 1 ft. from the outside, or an embankment made in rear 18 in. high.

Notwithstanding these precautions, walls may still give cover, hence they should be flanked when possible.

162.—In preparing walls for defense, the following cases arise :—

(1) A wall less than 4 ft. high. Sink a small trench on the inside to gain additional cover. Fire over the top. (Pl. 24, Fig. 1.) Head cover should be provided with logs, sandbags, or sods, sandbags being the best, sods next.

Additional protection against artillery may be obtained by heaping earth from the ditch in front against the wall, the thickness depending on the kind of artillery the wall is to resist.

If this should be done, the ditch should not be too close to the wall. A trench should be dug in rear to give cover to the supports, or for the firing line when not firing from the wall.

(2) A wall between 4 and 5 ft. can be used as it stands, subject to the same modifications as in the preceding case. (Fig. 2.)

(3) Between 5 and 6 ft. a wall can be notched. The tops of the notches may be filled with sandbags, sods, etc. (Fig. 3.)

(4) Should the wall be higher than 6 ft., a platform or staging must be raised inside to enable the men to fire over the wall or through the notches, or else the wall must be loop-holed.

163.—Loop-Holing. Loop holes for fire should not be closer than 2 ft. 6 in.; ordinarily they should be 3 ft. To find the height *hold the rifle in the position intended to be used.*

164.—To make a loop hole in a wall (Fig. 4), 14 in. or less in thickness, begin on the *inside*, to prevent the splay being toward the outside, by detaching a stretcher: the adjacent header on the outside can then be knocked out and the loop hole roughly shaped. Outside dimensions, 4 in. wide by 3 in. high. Interior dimensions will depend on the nature of the ground over which fire is to be delivered. For horizontal fire increase the breadth, for elevated or depressed fire increase the height.

In walls of ordinary thickness, a loop hole can be made in about 15 minutes, a notch in about 5 minutes.

165.—For a thick wall (Fig. 4), the small part should be at the center, the loop hole splayed to the front and rear, for it will do this anyway. The side toward the enemy may be stepped, in order that bullets striking it may flatten. Loop holes of observation should splay towards the outside.

All loop holes, when not in use, should be blinded.

166.—The height or position of the loop hole is influenced as follows:

(1) 1 ft. above the ground. Men lying down in a shallow excavation. Earth heaped up to 18 in. in rear prevents the enemy from using the loop holes should he close on the wall. The loop holes are a difficult mark for the enemy and sentries at night can watch the sky line. Cannot be used where ground in front is broken.

(2) Loop holes 2 ft. 3 in. Sitting. Position easy, but must have a deep ditch in front.

(3) Loop holes 3 ft. Kneeling. Position strained and must have a deep ditch in front.

(4) Loop holes 4 ft. 4 in. Standing. Good command, but easier mark for enemy: ditch necessary, but not so deep as in cases 2 or 3.

(5) Loop holes 6 ft. or more. Men standing on banquette. Best position for view or fire; no ditch in front; banquette strengthens wall, but takes longer to prepare wall in this manner; more susceptible to artillery fire.

167.—In order to allow a double tier of fire, walls should be at least 9 ft. high. (Figs. 5 and 6.)

PLATE 25.

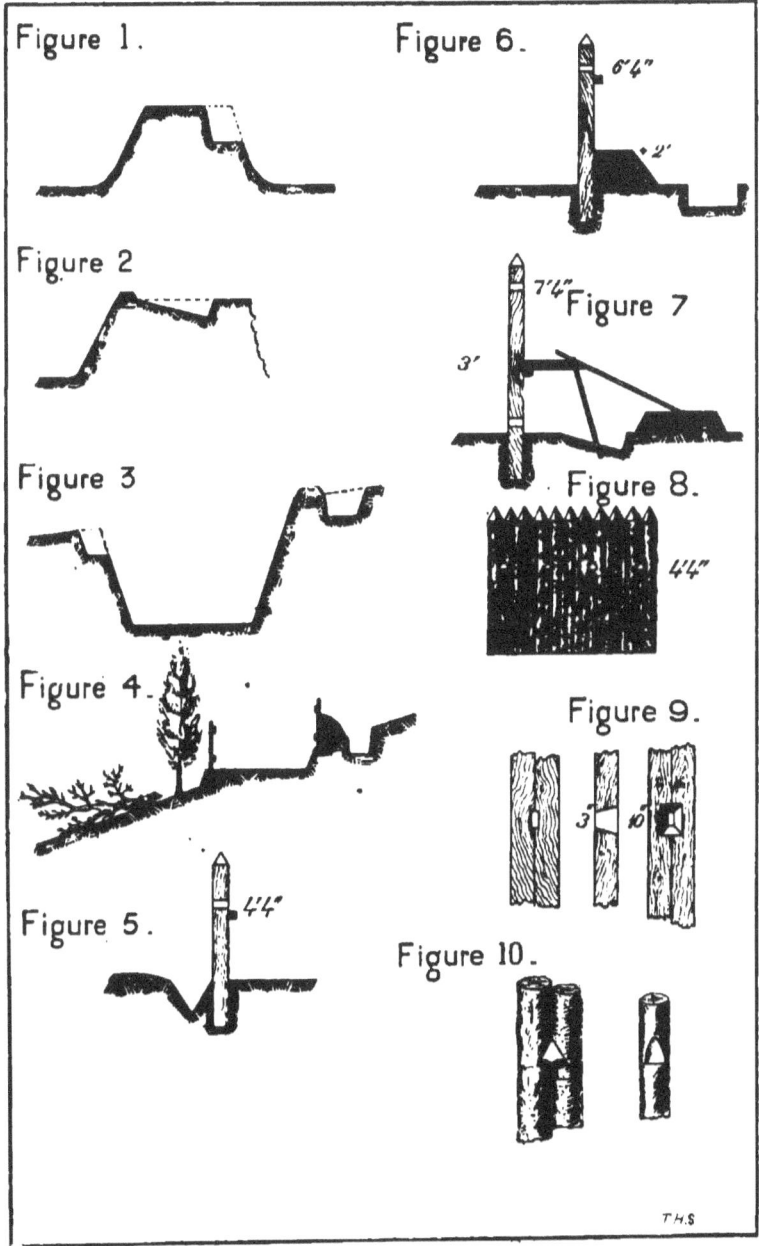

DEFENSE OF LOCALITIES. 113

168. Fences. Fences should be removed or left standing according to their position or direction. Wire fence forms a good obstacle, a rail or plank fence forms a screen and may be banked with earth to give cover. If fence is of stone it may be treated as explained for walls.

169. Hedges. Almost the same principles as explained for walls apply to hedges. A hedge primarily acts as a screen, but to resist projectiles must be banked up with earth.

Hedges possess the following advantages: (1) With little labor afford good cover. (2) Serve as a screen and also as a revetment. (3) Act as an obstacle to the enemy.

To derive these advantages hedges can be treated as follows:—

(1) A hedge with a ditch on the defenders' side. Can be used as it stands, the ditch being converted into a trench, widened and improved if necessary. (Fig. 7.)

(2) A hedge with a ditch on the enemy's side. Deepen the ditch, if necessary, and throw the earth to the defenders' side to give cover. If this is not possible, scatter the earth and dig a trench in rear. (Figs. 8 and 9.)

(3) A hedge with ditches on both sides. Deepen the ditch on the enemy's side, using the earth to obtain cover. Trench on defenders' side may be deepened. (Fig. 9.)

(4) Hedge on sloping ground. Gain cover by a small trench in rear; scarp away the ground in front, forming a glacis. (Fig. 10.)

(5) High and strong hedges. When time is available, may be treated as in Fig. 11. This is advantageous where additional command is required.

(6) A sunken road with hedges on both sides. Dig a shelter trench in rear of the hedge on defenders' side and utilize the earth to form a breastwork. Cut down the hedge on the enemy's side, entangling it to form an obstacle.

(7) Hedge without ditches. Excavate a shelter trench on defenders' side and bank up the earth against the hedge as a breastwork.

Weak places in hedges should be closed up with boughs, stakes, wire, etc., as a strong hedge, in addition to being a screen, forms a very efficient obstacle.

170. Embankments may be defended:—

(1) Narrow Embankments. By occupying the inner edge better cover is obtained, but, unless the bank is both low and narrow,

there is a dead space in front at the foot of the outer slope. (Pl. 25, Fig. 1.)

(2) Broad Embankments. By occupying the front edge, a better field of fire is obtained, but less cover can be provided for the firing line, and the supports are exposed when coming into action. (Fig. 2.)

171.—Cuttings are usually defended on the defenders' side, since in this case retreat is easy and the cutting itself forms an obstacle to the enemy's advance. But for active defense the front edge may be held and then a forward movement is possible. In this case a means of retreat must be provided.

172. Fig. 3 shows a case where the fall of the ground admits of both sides of the cutting being occupied, giving a double tier of fire. The command of the higher edge over the lower should be about 6 ft.

173.—In case of a road cut, as in Fig. 4, the upper fence may be used to sustain a breastwork, while the hedge below may be converted into an obstacle.

174. –Woods.—(Pl. 26, Fig. 6.) **Preparation of edge of wood occupied.** The edge of a wood should be put in a state of defense and an abatis is the readiest means of doing it. The salients should first be prepared, the re-entrants next and then roads entering the wood from the enemy's side. Instead of an abatis, the outer trees may be left standing and an entanglement made by packing in among them smaller trees cut about 10 to 20 paces to the rear. This clearing will serve as a communication all round the edge of the wood.

If a re-entrant bend is deep the abatis or entanglement may be carried straight across it and flanked from the adjacent salients. In case of a road, a lunette may be used instead of the abatis or entanglement. In case there is not time to prepare the entire edge of the wood, the salients may alone be prepared, the flanks of the defense being turned back for a short distance into the wood.

175. Preparation of woods lying beyond. Woods beyond, within rifle range of the line of defense, but too far to the front to be occupied, and too extended to be felled, should have an abatis or entanglement on the rear side to act as an obstacle.

176.—Cover. Trench digging is difficult on account of the roots, but when possible it should be done. Cover is generally

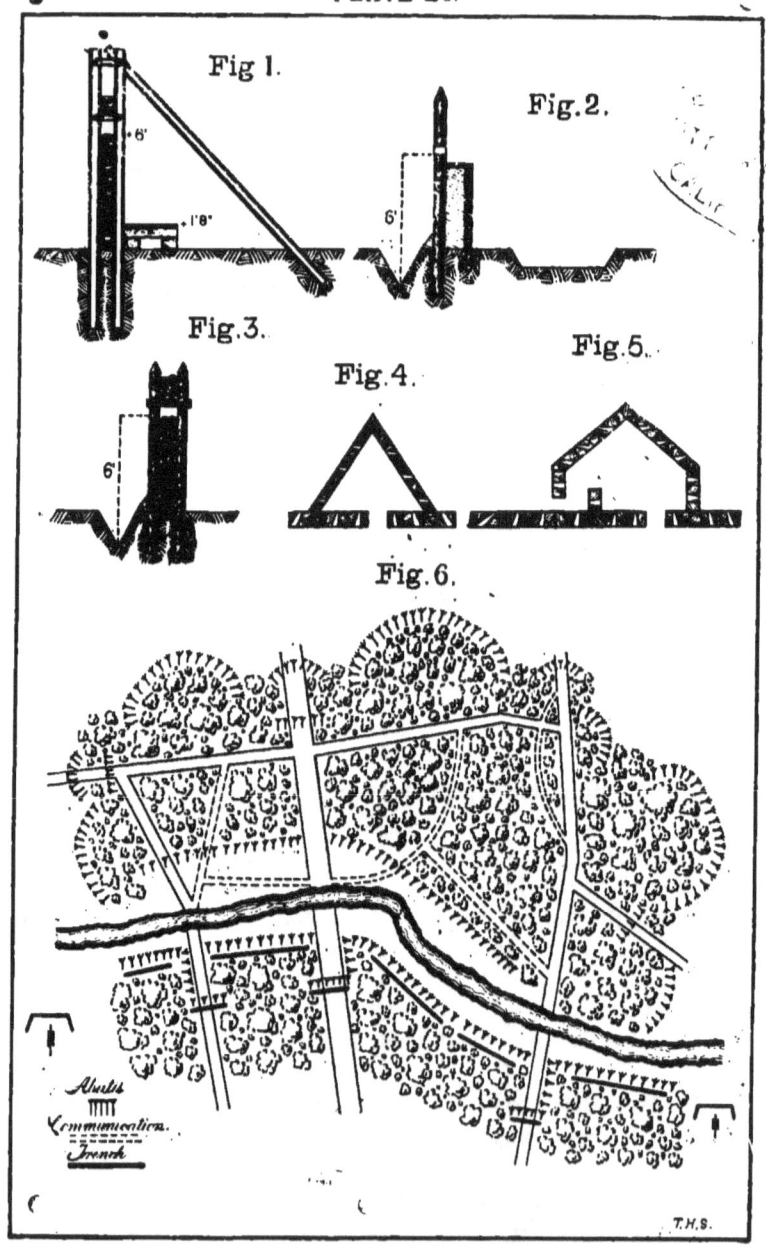

PLATE 26.

DEFENSE OF LOCALITIES. 117

obtained from the natural features. Trees, unless very large and standing thickly, do not give complete protection against artillery fire. Troops as supports and reserves, if not so far to the rear as to preclude their seeing through the wood to the front, may be covered by log walls and trenches.

177. Communications. (a) There must be good radial communication as well as free movement along the boundary in rear of the firing line.

(b) Roads and paths for bringing up the supports must be clearly marked by blazing as well as by posting sentries at all cross roads.

(c) In dense woods, preparations should be made for blocking up roads by cutting trees on either side of them nearly through, to be pulled down across them in case of retreat.

178.—Second and third lines may be placed along any open space, brook, or broad road, parallel to the front. In case of brook, the brook should be in front of defenders' position.

179.—Artillery should generally be placed outside of the wood on the flanks. If placed in the wood, batteries should be placed far apart, near good roads, masked as much as possible and each gun having more than one position. Re-entrants are desirable.

The number of defenders is estimated at 2 to 3 men per yard of front.

180.—Stockades. Stockades are timber defenses, made by placing one or more rows of timbers or rails, upright or horizontally, and so close to one another as to keep out rifle bullets, loop holes being made through which fire is delivered. They have the advantages of combining a parapet and an obstacle in one, giving good cover and ample interior space, and of being easily guarded against surprise. On the other hand, they require considerable time for construction, a certain amount of skilled labor, and are easily destroyed by artillery fire.

181. Stockades would be employed:
(1) Where timber is plentiful.
(2) When artillery fire is not to be resisted.
(3) When acting purely on the defensive. They are useful for the rear faces or gorges of enclosed works, and may be a good deal used in the defense of houses, streets, villages, and even woods.

182.—Stockades of vertical timbers. Vertical timbers

should be close together, planted in the ground to a depth of 3 or 4 ft., according to their size and weight, pointed or spiked at the top, and loop-holed at intervals. A riband must be spiked along the *inside*, near the top, to keep the timbers close together.

183.—Pl. 25, Figs. 5,6, and 7, show stockades with squared timbers; Fig. 8 with round timbers squared where they touch and the joint between every two trees made good on the inside by a smaller tree.

184.—The loop holes should be made in the crack between the timbers, in order to avoid weakening them, half being cut out of each. (Fig. 9.) In round timber, two saw cuts will make a loop hole. (Fig. 10.)

185.—The loop holes should be cut before the timbers are placed in position and the same precautions in regard to them as given in walls should be observed. A loop hole can be cut in from 10 to 15 minutes.

186.—In the foregoing cases, where the stockade is built of timbers placed vertically, squared timbers are preferred, as they are more easily fastened together and the joints made bullet proof. In round timbers the logs should be as straight as possible. If very crooked, two complete rows will be required.

One N. C. O. and 10 men will erect 15 running feet of stockade of squared timber, with one tier of loop holes, in 8 hours.

187.—**Stockades of horizontal timbers,** iron rails, fascines, or logs. (Pl. 26.)

Fig. 1 shows stockade of rails and ties. Can only be used for a very short distance as it will involve an immense amount of plant. Can be used to cover guns, close a road, and is more properly a barricade.

Fig. 2 shows a stockade or log breastwork, banked in rear with earth held in place by planks or hurdles and stakes.

Fig. 3 shows a stockade or breastwork of logs and fascines.

188. Stockade work, both vertical and horizontal, can be used for the construction of *tambours* (Figs. 4 and 5) and *caponiers* for flanking walls or stockades and covering entrances. Tambours may be triangular or rectangular in shape, arranged for one or two tiers of fire, and covered with a splinter-proof roof.

189.—**Buildings.** Buildings may be used for defense, either singly or in combination:

(a) As tactical points in the battlefield, held either as advanced posts or as supporting points in the line, or on the flanks, or as rallying points to cover retreat.

(b) As keeps to a more extensive position, such as a wood, village, etc.

(c) As an isolated post on the lines of communication.

190.—In order to admit of use as a defensible post, a building should possess the following requisites:—

(1) Solidly built of soft stone, brick, or adobe.

(2) Large enough to hold at least half a company.

(3) Sheltered from distant artillery fire, otherwise the building cannot be held against infantry or cavalry.

(4) Well selected for the object in view.

(5) Low, flat roof.

(6) Clear field of fire obtainable.

(7) Shape in plan affording flank defense.

191.—The building should be looked upon as a *keep*, or *second line* of defense, a first line being prepared at a *minimum* distance of 40 yds. to the front, this distance being the least that will give the defenders immunity from splinters caused by shells striking the building.

192.—In falling back, the first line should retreat *past*, not into the house, which should by this time be occupied by the supports. The garrison of the house may be estimated at two men to each door, window, or loop hole, with a reserve of one-fourth, tactical unity being in this, as in all similar cases, adhered to as much as possible.

193.—The following are the steps which must be taken in hastily preparing a house for defense:—

(a) Remove the inhabitants, also all easily combustible material, and provide water and heaps of earth in each room.

(b) Barricade doors and ground-floor windows (bullet proof if possible), also mask inaccessible windows, and remove all glass.

(c) Make loop holes in doors, shutters and walls, and, in the case of a sloping roof, remove tiles or slates.

(d) Clear away cover in the vicinity as far as time and means will allow.

(e) Open up communication throughout and prepare a means of retreat.

194.—The same precautions as to loop-holing walls apply in case of buildings. On the ground floor the horizontal dimension of a loop hole should be greatest; on upper floors, the vertical dimension. If an artillery attack is feared, shelter trenches should be provided outside the building on the flanks.

195.—Barricades for Doors may be made in the following ways: -

(a) Fill boxes, barrels, cupboards, etc., with earth and place them against the door *inside*.

(b) Build a wall of brick, stone, flag-stones, or hearth-stones, against the door inside, and support by a shutter or another door.

(c) If railway plant is available, pile ties horizontally on one another and secure with telegraph wire.

(d) Pile lumber inside the door and fix with blocks nailed to the floor.

(e) Other methods may be employed in accordance with material available.

196.—Should a door be reserved for use, it should be in a re-entering angle of the building, if possible, and protected from fire. A couple of chests filled with earth and placed on rollers may be used to secure the door. Similarly it may be possible to place iron or wood on the door, thus rendering it bullet proof.

197—Windows. Windows must be barricaded as explained for doors. If provided with shutters, these should be utilized. Upper windows require to be bullet proof only high enough to cover the defenders. Bedding is no protection against modern rifles, but may be used to mask windows of upper floors. If timber is used it should be placed vertically and nailed to horizontal ribands strutted back to the floor.

198.—If the house is large and strong and is to be held to the last, in addition to the foregoing, the following preparations should be made: –

(1) Arrange for storage of provisions and ammunition.

(2) Set apart a place for a hospital.

(3) Prepare latrines.

(4) Loophole partition walls and upper floors.

(5) Make ready barricades to cover retreat from one part of the building to another.

(6) If artillery is feared, shore up the floors and cover them with about 3 in. of earth.

199.—Should the construction of the house not afford sufficient flank defense, it can be improvised in the shape of tambours

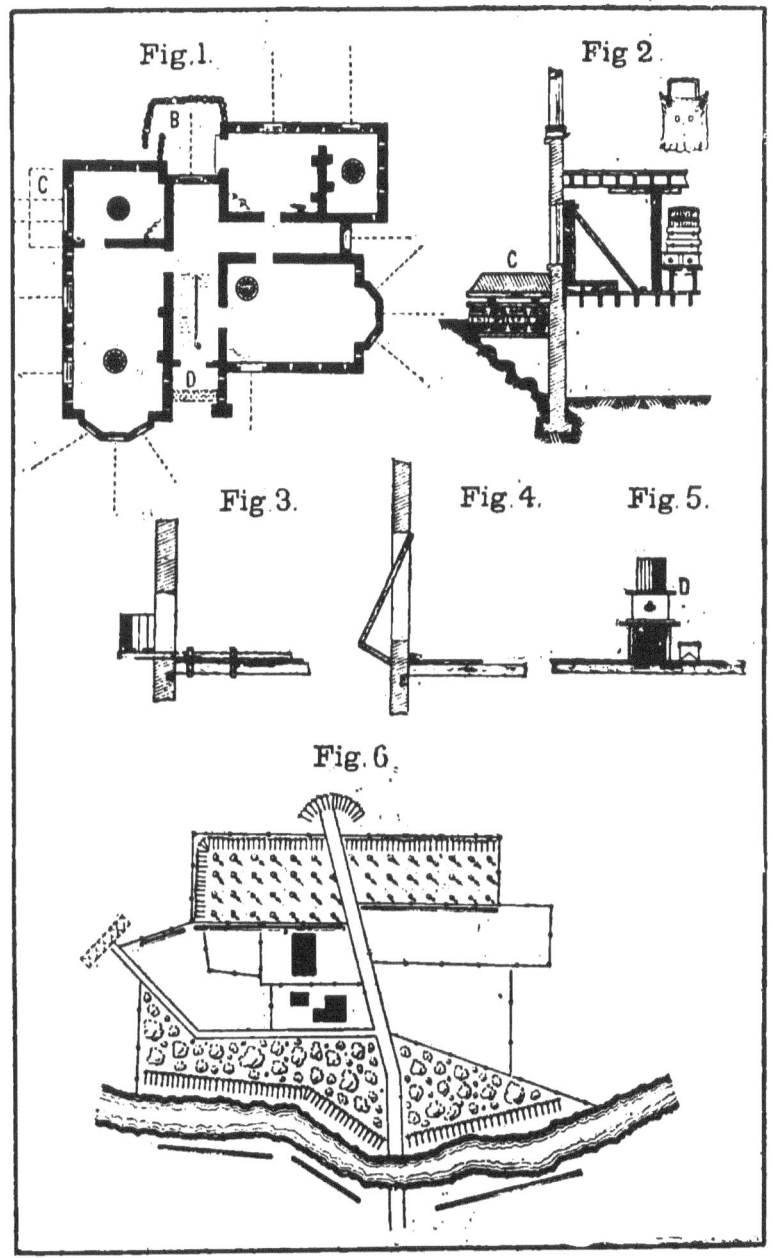

or caponiers, but the labor involved in their construction is considerable and they would only be undertaken for the defense of a very long wall or to cover an important entrance or communication.

For the latter purpose a *machicoulis gallery* is sometimes employed. (Pl. 27, Fig. 3.) This is made by removing the wall of the upper story where a window occurs down to the level of the floor, running out two or three long balks so as to project a few feet beyond the wall, the other ends being secured down to the floor. On these planks are nailed, with holes cut through to act as loop holes, and a musket-proof parapet of planking, sand-bags etc., is built all around. A projecting veranda offers a favorable position for this arrangement.

Second method: If a regular gallery can not be made, holes may be cut in the wall at a convenient height for a man to fire downwards when leaning over, and a screen of wood or other material may be secured outside for protection. (Fig. 4.)

If neither of the foregoing methods be possible, holes may be made in the roof, through which grenades may be thrown on the enemy.

200.—The materials most likely to be useful in preparing a house for defense are sand-bags, stout timbers, such as railway ties, large boxes, chests, barrels, coal-boxes, furniture and bedding.

201.—Pl. 27, Figs. 1, 2 and 5, illustrate the more important points in the defense of a house.

202.—Farms. Farms should be defended according to the nature of the surface covering, the ground and the improvements, and may involve the preparation for defense of walls, hedges, cuttings, embankments, buildings, woods, etc. Owing to their positions, farms may become very important and a great amount of fighting take place for their possession. They may occur either in the main line of a position, as an advanced post in front, or as a reserve station or rallying point in rear.

203.—Fig. 6 shows the principles of defense applied to a farm lying in advance of a stream, which is a point that requires to be strongly held. From the position of the farm it must be held as an advanced post.

The firing line is established along the fences bounding the fields and orchard. The farm buildings are loop-holed and can be held should the firing line be forced, while the fire from the house would render occupation of the farmyard by the enemy difficult. Further to the rear, the wood is strongly prepared for a final position, as shown in the figure.

204. The rear of an advanced post should be left weak and open to facilitate recapture.

205.—**Villages.** Villages can be rapidly prepared for defense and, under favorable circumstances, obstinately defended; consequently they are valuable supporting points in a defensive line. Owing to the effect of modern artillery and the ability of bursting shells to set villages on fire, great precautions have to be taken in the preparation for defense.

206.—A village, when properly prepared and defended, may have the following advantages:—

(a) Can be rapidly placed in condition for defense.
(b) Defense may be obstinate—thus giving time.
(c) Conceals the strength of the defenders.
(d) Provides a certain amount of cover from fire.
(e) Shelter from the elements.

On the other hand:—

(a) The garrison is scattered, hence the difficulty of supervision.
(b) When under artillery fire, splinters may cause many casualties.
(c) Liability to be set on fire by shells.

207. —A village may be held with the following objects in view:—

(a) As a supporting point in the main line of defense.
(b) As an advanced post in front of the main line.
(c) As an independent post.
(d) As a reserve station or rallying point in rear.

In the *first* case, strengthen the front and flanks. The rear should be prepared to resist infantry. In the *second* case, the distance from the main line will govern the amount of preparation. If very distant, should be prepared for all-round defense. If within rifle range, the rear should be left open, so that in case the village is taken, recapture will be facilitated. In the *third* case, if an independent post, must be prepared for an all-round defense. In the *fourth* case, if in the rear of the main line, must be prepared for a protracted, all-round defense.

PLATE 28

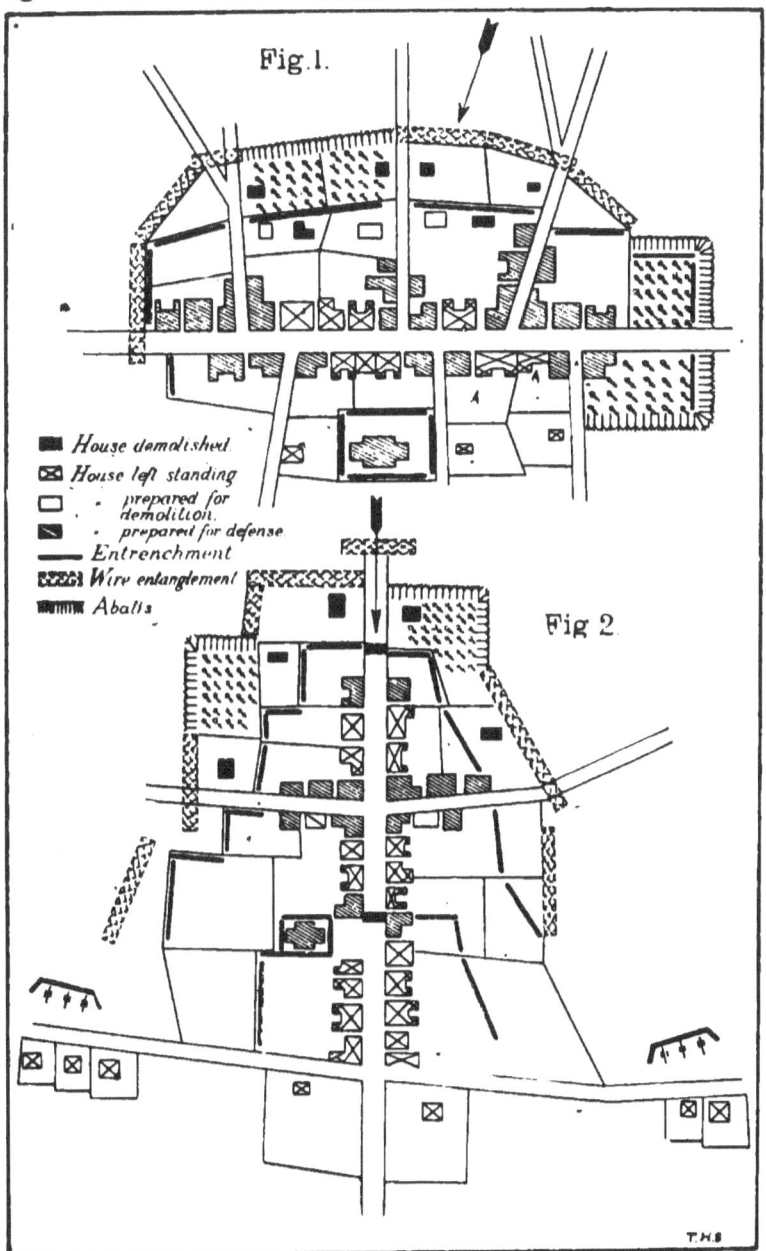

DEFENSE OF LOCALITIES.

208. Whether or not a village is to be held will depend on:—

(1) Its tactical value as compared with the number of men required to defend it.

(2) Whether it is practicable to provide a sufficient garrison for its defense.

(3) Whether it will be possible to demolish the village entirely, in order to deprive the enemy of the cover it provides.

(4) On the form and nature of the surrounding country, *i. e.*, no commanding ground within artillery range, foreground easily prepared and the unimpeded advance of the defenders' troops in the required direction easily arranged.

(5) On the shape of the village—whether broadside, salient, or circular.

(6) Nature and materials of the houses.

209.—The first points to determine in preparing a village for defense are how much of it will be defended, whether there are buildings suitable for a keep or citadel, and whether or not these are properly located.

210. The arrangements for defense would be made in the following order:—

(1) Clear the ground toward the enemy. (See Chap. V.)

(2) Cover for the firing line, supports, and reserves. (See Chap. IV.)

(3) Creating obstacles. (See Chap. VI.)

(4) Preparing communications. (See Chap. XVII.)

(5) Constructing retrenchments, citadels, or keeps. (See "Buildings.")

211.—The garrison of a village may be estimated at two men to the yard of perimeter to be defended.

212.—Salient Village. (Pl. 28, Fig. 2.) The successive lines of defense must be carried well out to both sides and the flanks well protected, otherwise the enemy may turn them and avoid fighting in the streets.

213. Broadside Village. (Fig. 1.) Here the outside fences must be more utilized than the actual buildings, as the latter are open to fire from artillery.

214.—Circular Village. (Pl. 29.) Great attention must be paid to the proper division of the village into sections for defense and preparing and making the communications.

215. In any of the foregoing cases, if cover does not exist for supports and reserves, it must be provided, as the village will probably be shelled before being assaulted.

If artillery is to be used it should be placed on commanding ground, inaccessible, if possible, to the enemy, and so that its fire will sweep those parts most favorable to the enemy's advance.

PLATE 29.

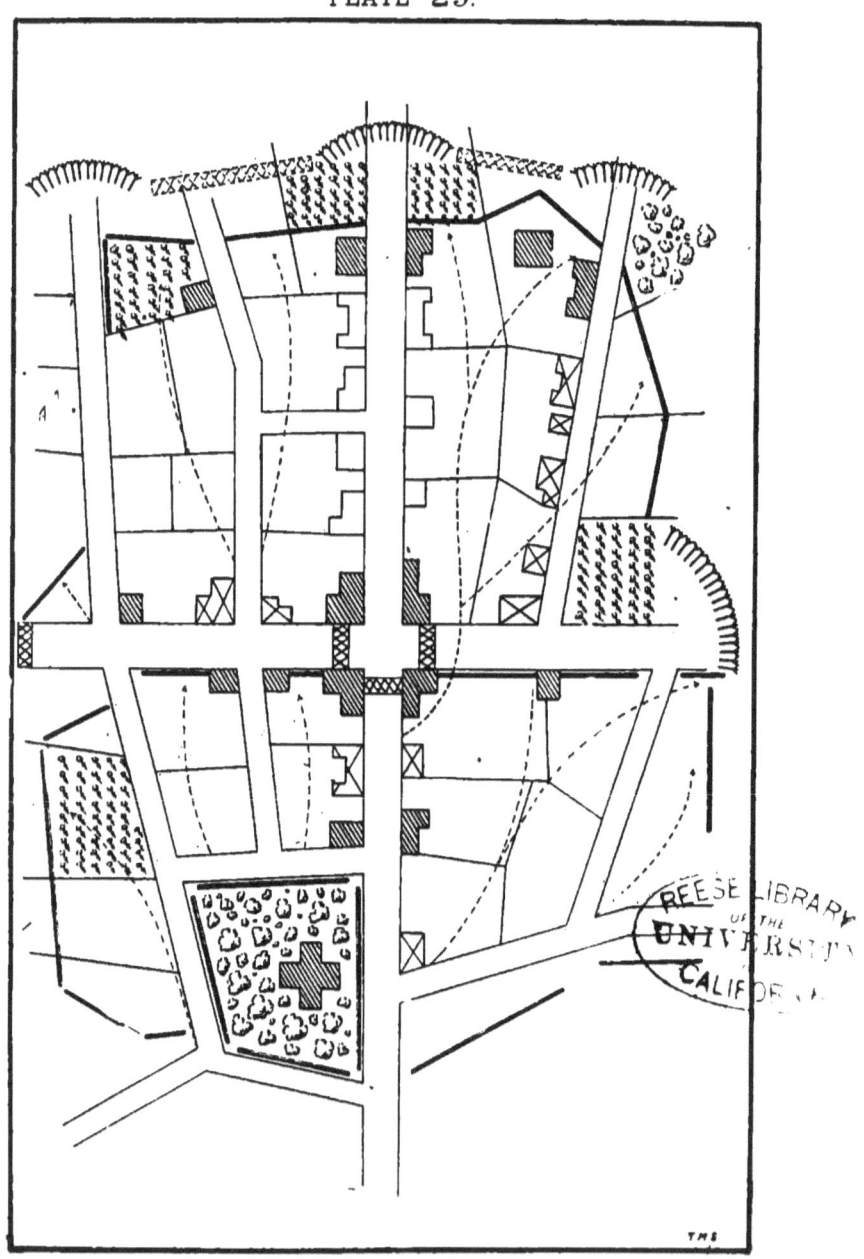

CHAPTER XIV. Use of Cordage and Spars.

216.—A rope is composed of three or more strands of fibrous material, iron or steel, twisted together. The strands of fibrous ropes are formed of threads; of iron and steel ropes, of wires. The size of rope is denoted by its diameter in inches,* and is generally sold by the pound. Fibrous ropes when new and dry stretch considerably, when wet they contract; advantage is often taken of the latter fact to tighten temporary lashings. Manila rope is only about $\frac{2}{3}$ as strong as hemp rope; tarred ropes only about $\frac{3}{4}$ as strong as untarred.

217.—*A rule approximating to the breaking weight* of a new rope, in tons of 2,000 lbs., is to take one-fourth the square of the circumference in inches. The strength of pieces from the same coil may vary 25 per ct.

Ropes in daily use should not be worked up to greater than $\frac{1}{5}$ their breaking loads, to meet the reduction in strength by wear and exposure.

218.—The following table gives the approximate breaking loads and weights of new Manila ropes, Swede's hemp center Iron pliable ropes of 6 strands of 19 wires each, and hemp center Steel pliable ropes of 6 strands of 19 wires each, Manufacturers' Tests:

---

* In the Navy the size of rope is denoted by its circumference in inches. The method used should be distinctly stated.

USE OF CORDAGE AND SPARS.

Diam. in inches	Breaking loads in lbs.			Weight per 100 ft. in lbs.		Minimum Size of Sheaves in feet for Iron and Steel
	Manila	Iron	Steel	Manila	Iron & Steel	
1-4	780	....	....	3	...	...
3-8	1,280	5,000	....	5	26	1
7-16	1,562	6,200	12,000	6⅛	29	1½
1-2	2,250	7,600	15,000	8	35	2
5-8	4,000	11,000	24,000	13.5	70	2¾
3-4	5,000	17,500	36,000	16.5	88	3¼
7-8	7,500	23,000	50,000	24	120	3½
1	9,000	32,000	66,000	30	158	4
1¼	14,000	54,000	104,000	45	250	5
1½	20,250	78,000	154,000	66	365	6½
1¾	30,250	108,000	212,000	97	525	7½
2	36,000	130,000	250,000	115	630	9

219.—**Knots, Hitches, etc.** *The standing part* of a rope is any part not an end.

A bight is a loop formed in a rope. (Pl. 30, Fig. 1.)

Whipping is securing the end of a rope with twine to prevent it from fraying out. (Fig. 1.)

Parceling is wrapping a rope to prevent chafing or cutting against a rough surface or sharp edge. (Fig. 1.)

Stopping or seizing is fastening two parts of a rope together without a crossing or riding. (Figs. 1 and 17.)

Nippering is taking turns crosswise between the parts to jam them. (Fig. 1.)

Splicing is joining the ends of ropes by opening the strands and placing them inot one another (Figs. 2 and 3), or by putting the strands of the ends of a rope between those of the standing part. (Fig. 4.)

Rolling or stopper hitch for fastening a rope to a strap or tail block, and to secure a fall while being shifted on a windlass or capstan. (Fig. 5.)

Overhand knot to prevent the end of a rope from fraying out, from slipping through a block, and the beginning of several other knots. (Fig. 6.)

Figure of 8 knot, used in making cask piers. (Fig. 7.)

PLATE 30.

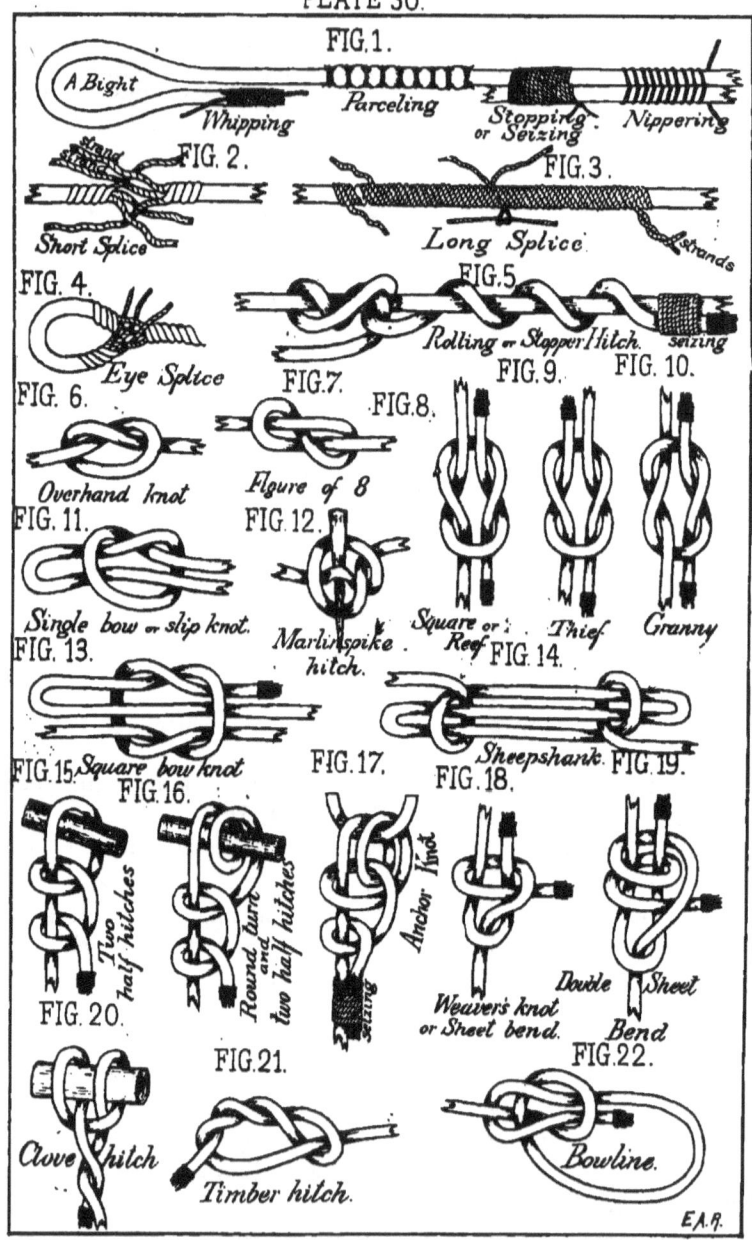

10

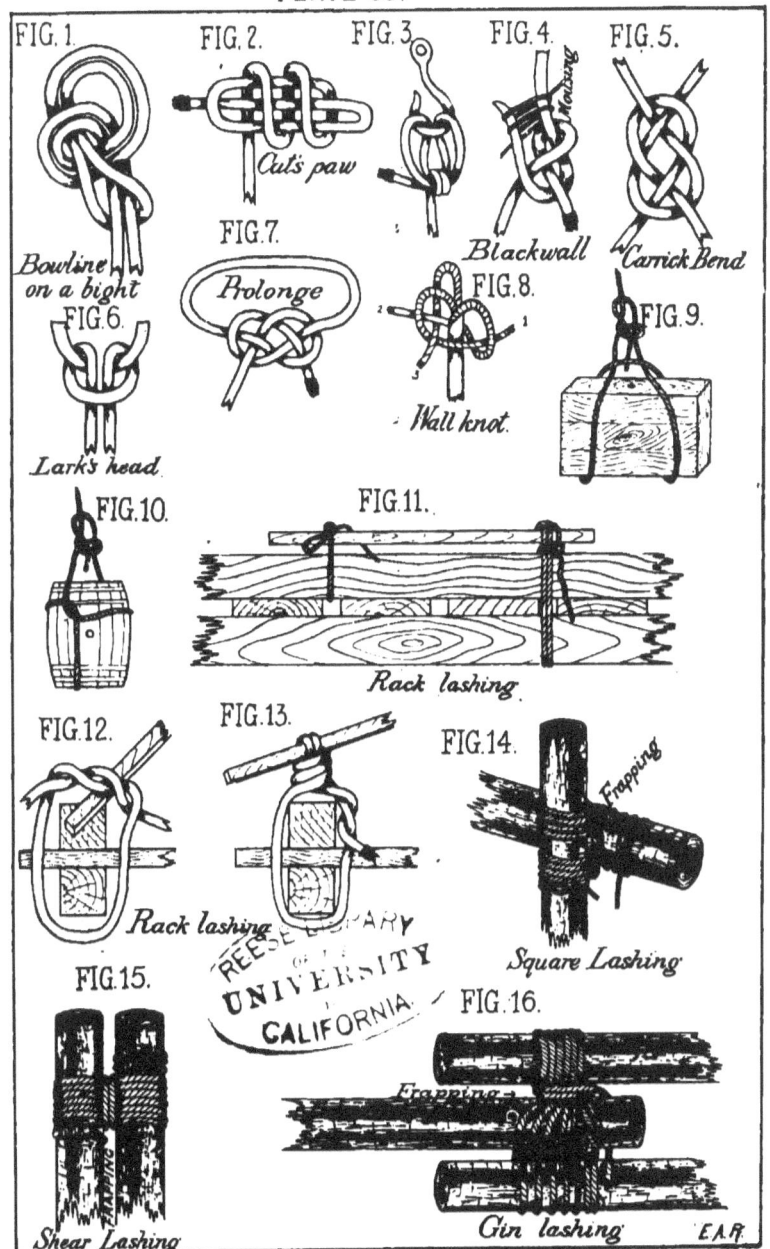

Pl. 31, Fig. 7, reverse rope in the three left and the two extreme right

Square or *reef knot* for joining the ends of two ropes the same size. (Fig. 8.)

Thief knot (Fig. 9), with ends on opposite sides, and *Granny knot* (Fig. 10), by crossing the ends the wrong way, both looking like square knots, are to be avoided, as they will not hold.

Single bow or *slip knot*. (Fig. 11.)

Square bow, which can be cast off. (Fig. 13.)

Marlinspike hitch, used in putting on lashings, etc. (Fig. 12.)

Sheepshank, used to shorten a rope temporarily without cutting. (Fig. 14.)

Two half hitches, for fastening the end of a rope around its own standing part. (Fig. 15.)

Round turn and two half hitches, to secure guys to stakes, etc. (Fig. 16.)

Fisherman's bend or *Anchor knot*, for fastening a rope to an anchor or ring. (Fig. 17.)

Weaver's knot or *sheet bend*, for joining ropes of different sizes without jamming. (Fig. 18.)

Double sheet bend, more secure than the single bend. (Fig. 19.)

Clove hitch, for fastening a rope to a spar; the end may afterwards be stoppered to its own part. The clove hitch differs from two half hitches only in being made around a spar or other rope instead of around its own standing part. (Fig. 20.)

Timber hitch jams when made round a timber. (Fig. 21.)

Bowline, to form a temporary loop at the end of a rope. (Fig. 22.)

Bowline on a bight, to make a loop on a bight. (Pl. 31. Fig. 1.)

Cat's paw, for applying a purchase or tackle to the fall of another. (Fig 2, the beginning; Fig. 3, how applied.)

Blackwall hitch, for fastening the end of a rope on a block in the simplest manner, or fastening a rope in a hook. (Fig. 4.)

Mousing is a seizing placed around a hook to prevent it from spreading or unhooking. (Fig. 4.)

Carrick bend, to fasten guys to a derrick. (Fig. 5.)

Lark's head, for fastening a bight to a ring. (Fig. 6.)

Capstan or *Prolonge*, making fast a spar. (Fig. 7.)

Wall knot, for finishing off the end of a rope to keep from unstranding (Fig. 8), by passing the strands, as shown, then drawing them down into a knot.

Frapping is passing a rope around a lashing to keep the turns together. (Figs. 14, 15 and 16.)

Straps are rings used for attaching tackles to spars or ropes. (Pl. 33; Figs. 1, 2 and 6.)

220. To make a short splice. (Pl. 30, Fig. 2.) Unlay strands of each end for a convenient length; take an end in each hand, place end to end, strands sandwiching, and grasp the three strands from opposite rope in left hand. Take a free strand, pass it over the first strand next to it, then through under the second and out between the second and third from it, then haul taut. Pass each of the remaining six strands in same manner, first those of one end and then those of the other, and so continue as far as desired.

221. To make a long splice. (Fig. 3.) Unlay strands of each end, 3 or 4 times longer than for short splice, and place end to end as described. Unlay one strand a considerable distance and fill up its space with opposite strand from other rope, and twist them together. Do the same with two strands on other rope. Open remaining strands, divide in two, make overhand knot with opposite halves, and lead ends as in short splice. Cut off the other two halves. Do the same with the other pairs of strands where twisted together. Before cutting off any of the half strands, first stretch, roll under the feet, and pound the rope well. This splicing does not increase the size of the rope and is used where the splice is to run through blocks.

222. To make an eye-splice. (Fig. 4.) Unlay one end for short distance, lay strands upon the standing part so as to form the desired sized eye. Put first end through the strand next to it. Put second over that strand and through second. Put third through third strand on other side of rope and so continue. This forms a permanent loop in end of rope.

223. To sling a box or barrel. Lay a long strap under it, spreading the parts, and pass one bight through the other; or, make a long loop with a bowline and sling as shown on Pl. 31, Fig. 9. If one head is out stand barrel up, put one part of a strap under middle of bottom, take a half hitch over top with each part just over bilge hoops and exactly opposite; or place rope under barrel, bring up over top. make overhand knot, open it out and slip each half down over hoops, fasten end to standing part with bowline. (Fig. 10.)

PLATE 32.

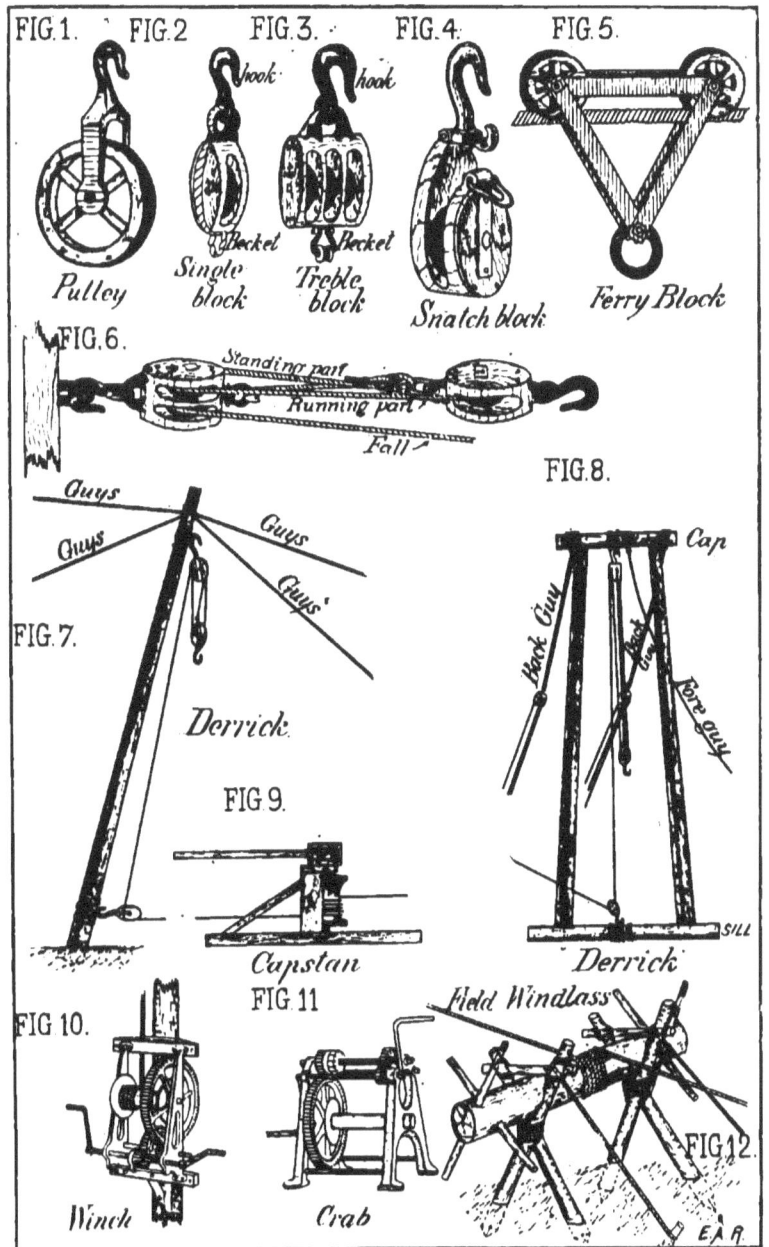

224.—Rack lashings (Figs. 11, 12 and 13) are made with a ⅝ in. rope, 18 ft. long with a loop at one end, and a *rack stick* 2 ft. long, 1¼ in. in diameter having a cord 4 ft. long through one end, by passing the rope two or three times around the side rail and balk, and, after making it fast, twisting it tightly with the rack stick.

225.—Transom lashing. (Fig. 14.) The spars are laid across each other at right angles, a clove hitch is made on one of the spars, the end then twisted around its standing part, then three or more turns are taken around the spars, under one and over the other, keeping outside previous turns on one spar and inside on the other. Several frapping turns are then taken between the spars and the end fastened on one of the spars with a clove hitch. Used in lashing transoms to standards in bridge building.

226.—Shear lashing. (Fig. 15.) The spars are laid parallel, a couple of inches apart, on a block, a clove hitch made on one spar, then 5 or 6 turns taken around both spars without riding. Several frapping turns are then taken between the spars and the end fastened on one of the spars with a clove hitch. This is used in rigging shears for hoisting heavy weights, etc.

227.—Gin lashing. (Fig. 16.) The three spars are laid parallel, a couple of inches apart, the butts of the two outside ones in one direction, that of the middle one in the opposite direction. A clove hitch is made on one spar, then 5 or 6 loose turns taken, passing over and under, without riding. Several frapping turns are taken in each interval and the end fastened on one of the spars with a clove hitch.

228.—Blocks, Tackles, etc. *A pulley* consists of a wheel, having a grooved rim for carrying a rope, turning in a frame. (Pl. 32, Fig. 1.)

A block (Figs. 2 and 3) consists of one or more grooved pulleys or sheaves turning on an axle, called a pin, mounted in a casing or shell, which is furnished with a hook, eye or strap on one end, by which the block may be attached to something and sometimes with a becket on the other end for attaching ropes, etc. It is used to transmit power, or change direction of motion, by means of a rope or chain passing round the movable pulleys. Blocks are single, double, treble, or fourfold, according as the number of sheaves or pulleys is one, two, three or four. The size of blocks is expressed by the length of the shell in inches. A common style of Ferry Block is shown in Fig. 5.

A *snatch block* (Fig. 4) is a single block with a notch cut in one cheek so as to receive the standing part of a fall without the trouble of reeving and unreeving the whole.

A *running block* is one attached directly or indirectly to the object to be raised or moved; *a standing block* is one fixed to some permanent support.

229.—A *tackle* consists of ~~one~~ two or more blocks with a rope rove through them for use in hoisting.

230.—The parts of all ropes between the points of fastening and sheaves are called *standing parts;* the parts between the sheaves are called *running parts;* the part to which the power is applied is called *the fall.*

231.—*To overhaul* a tackle is to separate the blocks; *to round in* is to bring the blocks closer together.

A tackle is said to be *block and block* or *two blocks* when the entire fall is hauled through so the blocks are in contact.

232.—Before reeving a rope in a block, it should be stretched out its full length. Tackle should not be allowed to twist; to prevent it, insert a bar in the block or between the running parts and use it as a lever to hold straight. If allowed to make one complete turn with 2 single blocks, the friction will increase the resistance about 40%. Ropes should not be too large for blocks, the rule being, "*Small ropes and big blocks.*"

233.— **Power of Tackle.** Theoretically, the power necessary to just balance a weight, with a tackle of two blocks, is equal to the weight divided by the number of ropes at the running block.

234.—To produce motion, however, a greater power is required to overcome friction and stiffness of rope. It has been found by experiment that to do this about 10% of the theoretical power necessary to balance must be added to itself for each of the sheaves over which the rope passes, the blocks being in good condition and well oiled. If not in good condition and not well oiled, the per cent may be as high as 30 for each sheave.

235.—The formula $P = \frac{W \times 1.8}{R}$ is used to determine the power required to raise a weight with a simple tackle, in which P = the power required; W = the weight to be raised, S = the number

PLATE 33.

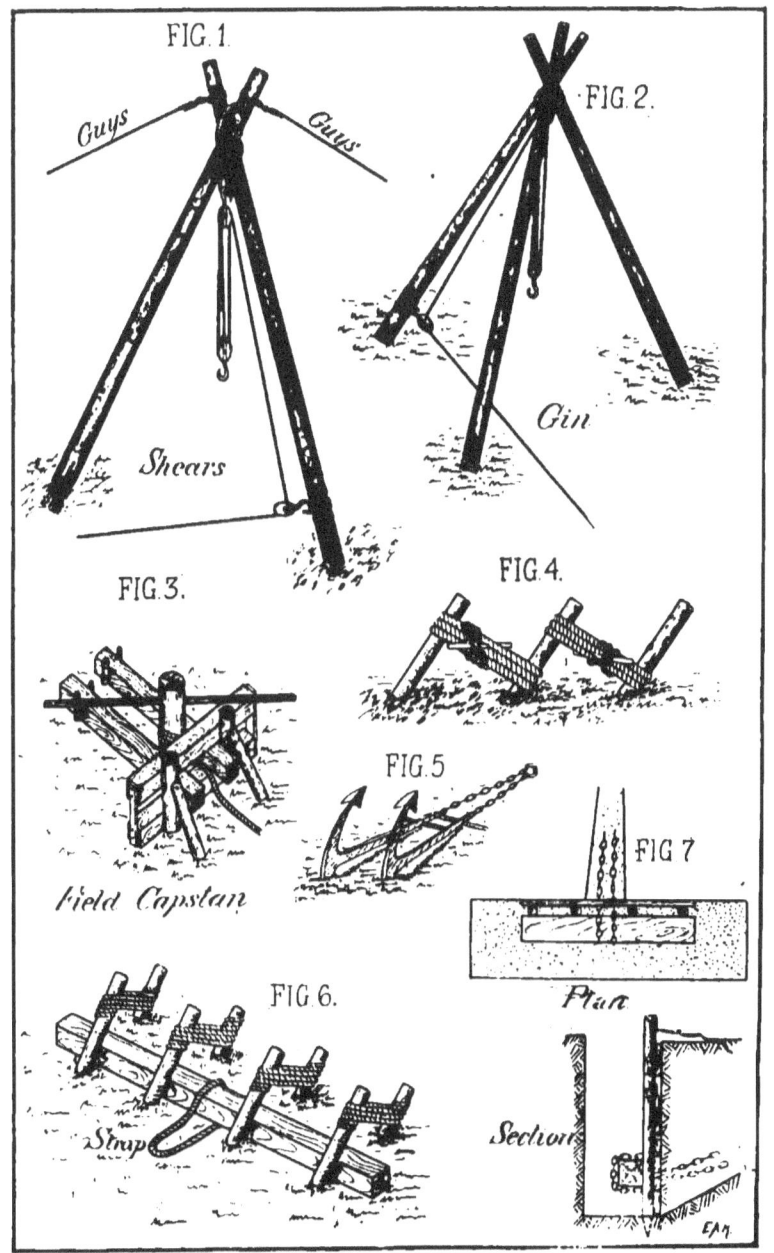

of sheaves, and R = the number of ropes at running block, including standing part if attached to it. If it is required to find how great a weight a certain power will lift the formula is $W = \frac{P \times R}{1.8}$. Power is gained only at the loss of time. The power moves as many times faster and farther than the weight as the number of ropes at the running block. No advantage is gained by using, in one fall, a greater number of sheaves than two treble blocks, but may be by a combination of blocks and tackles.

236.—A squad of men hauling on a fall exert a pull of about 80 lbs,, or half their weight each, the fall being nearly horizontal.

237.—A **Derrick** (Fig. 7) usually consists of a single spar or leg, held up by 4 guys, and having a tackle lashed to the top, used for hoisting or lowering heavy bodies within a circle whose diameter equals the height of the spar. When made of 2 legs (Fig. 8) they are mortised into a cap on top and a sill at the bottom, only two guys being required, a fore and back, but three are better, one fore and two back. The weight can only swing between the legs. The holdfasts for the guys should be at a distance from foot of derrick at least twice its height. The foot should be secured from slipping by being let into a hole in the ground or otherwise.

238.—**Shears** (Pl. 33, Fig. 1) consist of two spars, of a size suitable for the weight to be raised, lashed together at the cross.

A tackle is fastened at the lashing by a strap passed around it or otherwise, the hook moused, and holdfasts are required as for two-legged derrick.

Derricks and shears should not lean to exceed ⅙ of their height and each leg should have about ½ this inclination.

239.—A **Gin** (Fig. 2) is a tripod formed of three poles. The two outside ones are called legs, the third one the *pry pole*. Gins require no guys. Weights can only be lifted vertically.

240. In using derricks, shears, and gins, the fall is generally led through a snatch block lashed on a leg near the bottom, thence to a crab, windlass, or capstan. Derricks frequently have fastened on their legs *a winch* for transmitting the power. (Pl. 32, Fig. 10.)

241.—A **Windlass** (Fig. 12) consists of a horizontal axis fastened in a frame and turned by means of cranks or handles. The rope may either be fastened to the axis or passed two or three times around it, hauled taut, the free end being held, and taken in by men in the rear.

242.—A Capstan (Pl. 32, Fig. 9, and Pl. 33, Fig. 3,) consists of an upright barrel, either smooth or ribbed, arranged about a spindle. Above the barrel is the head with holes to receive the ends of levers or bars by which the barrel is revolved. The rope is passed and held as explained for a windlass.

243.—Holdfasts are stout wooden stakes driven into the ground, or other arrangements used for securing purposes.

An essential point to be considered before moving or suspending heavy weights is the nature and condition of the securing points, together with the strain that will be brought upon them. In the first instance, it is better to make them more secure than seems to be absolutely necessary, as, when they once begin to give way, it is difficult to strengthen them. Pl. 33, Figs. 4, 5, 6 and 7, show some of the various methods of making them, also Pl. 40a.

PLATE 34.

CHAPTER XV.—Spar Bridges.

244.—Military Bridges are not required to fulfil all the conditions of ordinary bridges. They are constructed for special and immediate purposes, usually with unskilled labor, and of such materials as can be procured on or near the spot. That the bridge built shall be strong enough to bear the heaviest load intended to be crossed is the first requisite; celerity and simplicity of construction next.

245.—Pl. 34 is an illustration of what was done in building Military Railroad Bridges under unfavorable circumstances in time of war with troops of the line, very few of whom were mechanics, many could not even handle an ax, none were trained to the duty, and none were engineer troops. This bridge was built by General Haupt over Potomac Creek, Va., during the Rebellion, and was 80 ft. high and 400 ft. long. It consisted of three tiers of trestles on top of cribs 12 ft. high. The timber used was chiefly round sticks, cut in the woods near by, and put together without bolts, simply with spikes and wooden pins, and, when finished, was crossed by 10 to 20 heavily loaded trains per day. This kind of work, however, properly belongs to a special construction corps, but it falls to the lot of the officers and men who first arrive at a stream on the ordinary roads, where there are no means of crossing, to construct an improvised bridge with such tools and of such materials as may be available.

246.—The plans and expedients which follow have been selected with a view to their being types of bridges that can be constructed by troops having no other tools than axes and augers, and such materials as growing trees found in the vicinity, and beams, boards, ropes, wire, nails, etc., obtained from neighboring houses and towns. The purposes for which the bridge is to be used, the nature of the crossing, velocity of stream, and kind of bottom, will determine its strength, kind, size, etc.

247. For a common road bridge, the load is assumed to be a maximum when covered with men, estimated at 120 lbs. to the sq. ft., plus the weight of the bridge, usually taken at about 80 lbs. per lineal ft. For reasons which are evident, the bridge should be as short as possible, with good approaches. Swampy, high, or steep banks should be avoided.

248.—Bridges usually take their names from some part of their construction, as *Trestle, Truss, Pile, Suspension,* or *Floating Bridges.* The distance between supports (determined by the strength of the balks to bear the desired load) is called *the bay* or *span,* and the corresponding part of the bridge *the span.* The *superstructure,* consisting of the stringers or balks, the floor, the side-rails and the fastenings, is of the same nature for each kind, as shown in Pl. 35, Fig. 1. The ends of the balks rest on cross pieces of the supports called transoms, on the balks (of which there are usually five) are laid chess or poles, forming the floor; on top of the floor, over the outside balks, are laid side-rails or poles, which are securely fastened every 4 or 5 ft. to the balks beneath by rack lashings. Hand-rails (Fig. 2) should always be provided on each side of the roadway. The usual width of military bridges is 9 ft. in the clear, between side-rails; 6 ft. will answer for Infantry in column of two's, and Cavalry by file; 2.5 ft. for Infantry in single file.

249.—For determining the strength of the materials to be used, all errors should be on the side of safety. The practical method is to place the ends of the timber on low supports, as far apart as they will be in bridge; as many men as can, then step on it and jump up and down; or it is otherwise arranged so as to bring as great a weight upon it as it will have to bear at any time in bridge.

Where small poles of the usual number would not be strong enough, a greater number must be used until the desired strength is gained.

250.—Transoms must be strong enough to bear all the weight that may be brought upon one bay of the bridge, considered as distributed dead load on the transom.

251.—*The load in pounds* which any timber resting on two points of support will safely bear, concentrated at its center, may be approximately determined by the formula $\frac{1}{3} \times \frac{bd^2}{l} \times C$, in

PLATE 35.

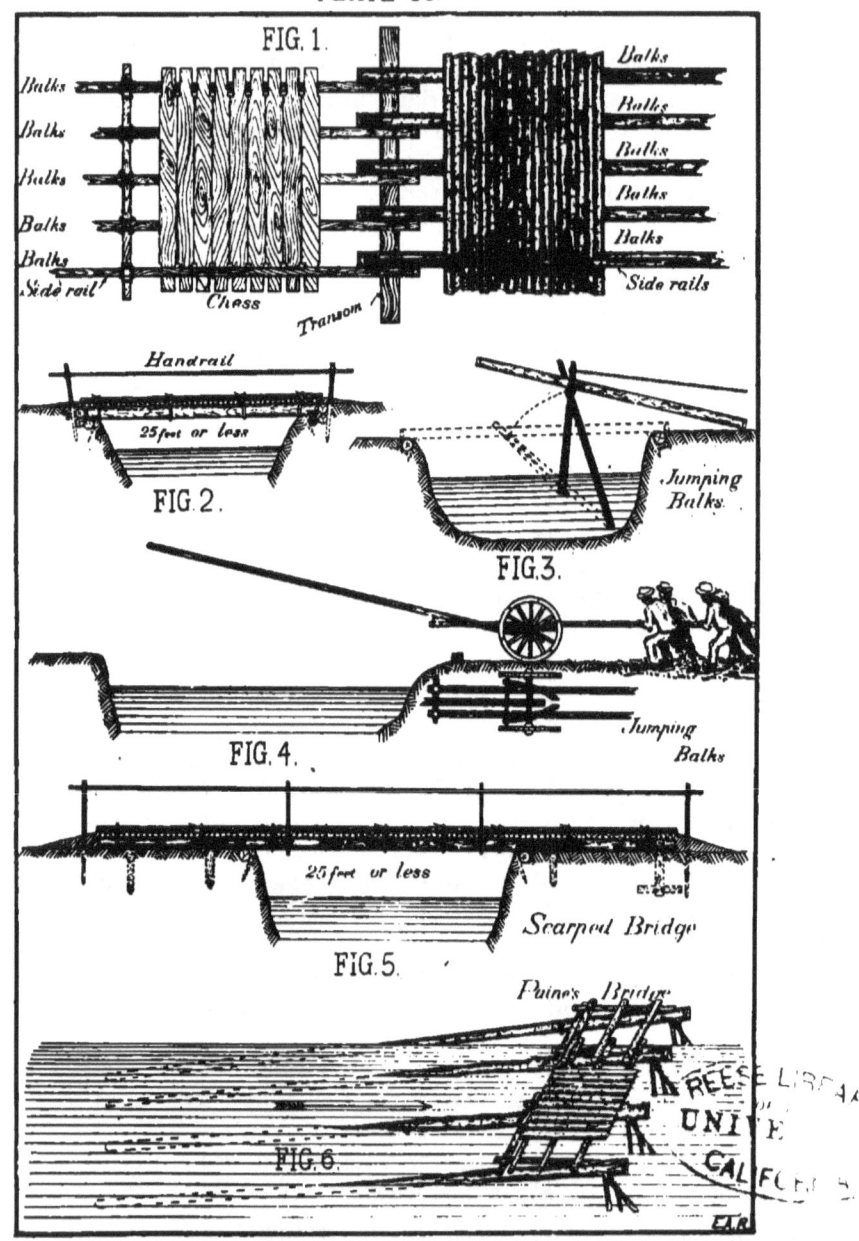

which b = the breadth in inches, d = the depth in inches, l = the length in feet between supports, and C is a constant in pounds for the particular material of the beam,* $\frac{1}{8}$ is the fraction of the breaking weight used for safety. It would, in all cases, be better to use a smaller fraction of the breaking weight, as $\frac{1}{5}$ or $\frac{1}{6}$; or even $\frac{1}{8}$ in cases of importance or of a permanent character. The formula is for a rectangular beam, but for a cylindrical timber whose mean diameter equals the side of a square beam, use $\frac{6}{10}$ of what the formula gives.

252. *Weight brought on a bridge by the passage of troops*, taken as distributed live load for Infantry and Cavalry: Infantry in column of twos or fours about 225 lbs. per lineal foot. Infantry when crowded at a check in fours about 550 lbs. per lineal foot. Cavalry in column of twos about 230 lbs. per lineal foot. Cavalry when crowded at a check about 350 lbs. per lineal foot. When Artillery carriages cross a bridge, the weight is not equally distributed, *but is greatest when the wheels bearing the heaviest load are on the center.*

253. A uniformly distributed dead load produces only one-half the strain of an equal dead load concentrated at the center. A moving or live load produces twice the strain of a dead load. A uniformly distributed live load equals a concentrated dead load.

TABLE OF CONSTANTS, C, for finding the breaking load of various materials by the formula $\frac{1}{8} \times \frac{bd^2}{l} \times C$ when concentrated at center of beam supported at both ends. From "Trautwine's Engineer Pocket Book:"

Ash, white....650 Elm.............350 Oak, red and black 550
Ash, swamp...400 Hemlock........400 Pine, white.......450
Ash, black....300 Hickory.........700 Pine, yellow......500
Beech, white..450 Hickory, pig nut..500 Pine, pitch.......550
Beech, red.....550 Locust...........600 Poplar............550
Birch.........450 Mahogany........450 Spruce...........450
Cedar.........250 Maple............550 Sycamore........500
Chestnut......450 Oak, white and live 600 Walnut..........450

*C is determined by taking a piece 1 in. square, and 1 ft. long between supports, loading it at the center until it breaks, then to the applied load adding one-half the weight of the piece between supports, and the sum will be C; or, a piece of any convenient size and length can be used, afterwards deducing what the breaking weight would be for a piece 1 in. square and 1 ft. long, remembering that the breaking weight varies directly with the width, as the square of the depth, and inversely as the length.

254. *The cubic contents* of a log is approximately equal to 0.7854 times the square of the mean diameter times the length, or the area of the mean section multiplied by the length; or the square of one-fifth the mean circumference times twice the length, all in feet.

255. In calculating the strength of a round timber or spar, its mean diameter is used, because such a spar, if overloaded, will break at center, instead of at small end.*

256. The following table gives the weights in lbs. per cu. ft. of various materials:

Iron, cast	450	Chestnut	40	Spruce	31
" wrought	487	Cottonwood	35	Sycamore	37
Lead	710	Hickory	43–49	Walnut	38
Steel	488	Maple	48	Clay	120
Ash	38–47	Oak	45–60	Earth	72–120
Cedar	35	Pine	34–40	Gravel & sand	90–130

Green timbers weigh from $\tfrac{1}{5}$ to $\tfrac{1}{4}$ more than those in table.

257. When spars are used for balks, they must be arranged so as to have all butts or all tips together on a transom. They should have good overlap and be well lashed to each other and the transoms. To allow for settling, the centre is generally made higher by about $\tfrac{1}{30}$ the span.

---

* For a bridge constructed as in Fig. 2, with 6 balks, to determine the safe load it will carry, the application will be about as follows:—

The balks being of yellow pine, 10 inches in diameter at the center, 25 ft. between supports, the formula gives $\tfrac{6}{10} \times \tfrac{1}{3} \times \tfrac{10^3}{25} \times 500 = 4{,}000$ lbs. as the safe load each balk will bear concentrated at the center, including its own weight. Calculating for 5 balks, on the supposition that the two outside ones receive only ½ the strain of the center ones, they will bear 5 x 4,000 lbs. = 20,000 lbs. concentrated at center. From this deduct half the weight of the balks and floor concentrated at center, found by multiplying ½ the cubic contents by the weight per lb ; for the 6 balks this will be

$$6 \times \tfrac{1}{2} \left[\left(\tfrac{5}{6}\right)^2 \times .7854 \right] \times 25 \times 40 = 1636.25 \text{ lbs.}$$

For the floor of 50, 4-in. poles, each 12 ft. long, the weight will be

$$50 \times \tfrac{1}{2} \times \left[\left(\tfrac{1}{3}\right)^2 \times .7854 \right] \times 12 \times 40 = 1047.24 \text{ lbs.}$$

20,000 lbs. − 2,683 5 lbs. = 17,316.5 lbs., the capacity of the bridge concentrated at the center. Infantry marching in column of fours crowded by a check would cause a load of only about 13,750 lbs. Cavalry in column of twos crowded by a check only about 8,750 lbs.

Similarly, knowing the span, the kind of material at hand, the weight to be borne, etc., the size of timbers required can be deduced; or, having the size and kind of the timbers, weight to be borne, etc., the greatest length that can be spanned can be determined by the above formula.

Foot note pg. 154, for "lb" read "ft." weight of floor deduced for 50 poles should be for 75.

PLATE 36.

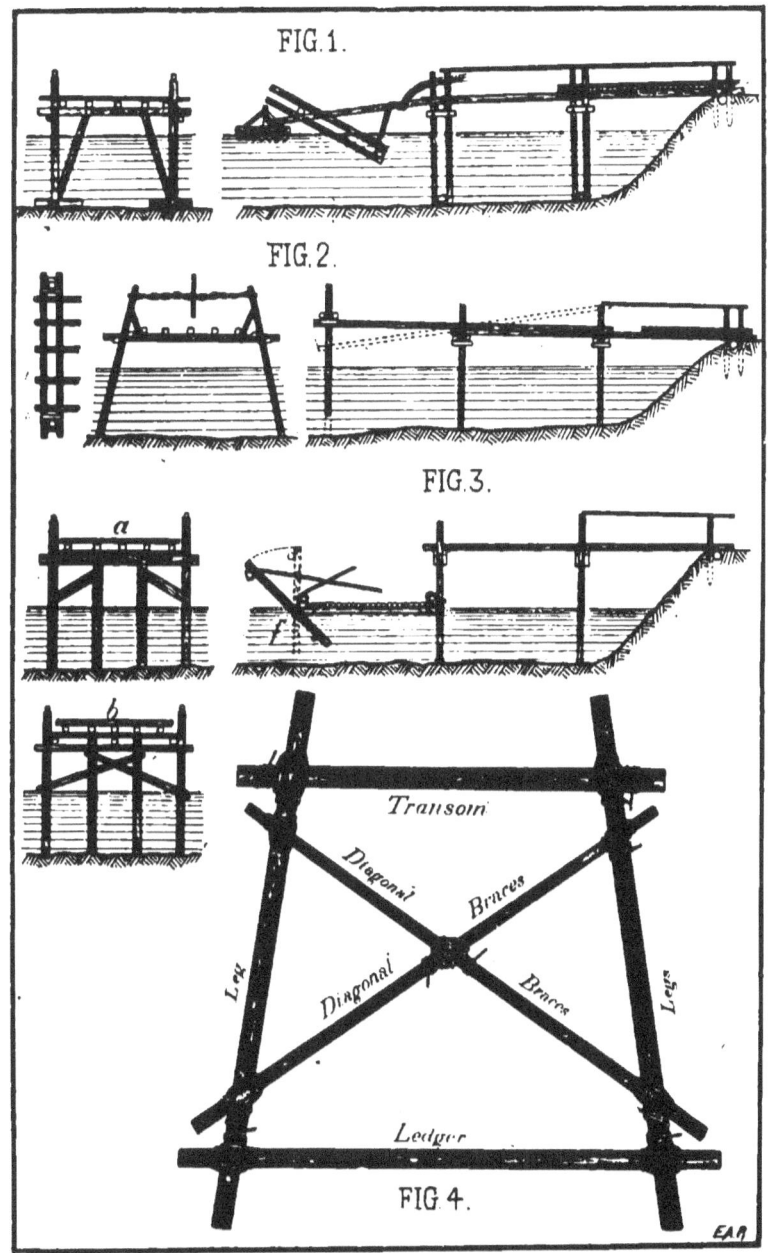

258.—*For spans of 25 feet or less*, if timber is available, the simplest form of stringer bridge could be built, as in Fig. 2, of 6 balks, 30 ft. long, reaching clear across, covered with small poles 4 to 6 in. in diameter, 12 ft. long, for a floor. Side-rails would be laid on the floor over the outside balks and either lashed or pinned to the balks. Hand-rails would be as shown or a rope stretched across would answer. Time of construction—1 hour. The balks could be jumped across as shown in either Figs. 3 or 4.

259.—If stringers of length to reach across cannot be obtained or are too heavy to handle easily, a bridge as in Fig. 5 might be made, requiring only axes and augers. The shore stringers, 25 ft. long, 10 in. in diameter, six on each shore, have their bridge ends scarped on upper side 18 in., then pushed out 10 ft., their shore ends being well anchored down and loaded with roadway. Six short stringers, 8 ft. long, 10 in. in diameter, three on each shore, scarped 18 in. on under side of each end are passed over gap and laid on shore stringers. Two 2-in. auger holes are bored at each end through both stringers and wooden pins driven through, and the flooring completed.

260.—Paine's Bridge. (Fig. 6.) If timber is abundant and stream not over 6 ft. deep, select trees up stream. Fell, and trim off branches. Bore two 3-in. auger holes near butt ends 3 in. apart, making an angle of 30° with each other, and a third hole, making an angle of 45°, between them nearer the butt. Cut and insert in outside holes legs long enough to raise the butt the desired height of bridge. Float down stream, butt end first, to position of bridge. On arriving in line of bridge, the log is turned on its feet, the tip sinking to bottom. The brace leg is then inserted down stream in last hole making an angle of 45°. Log after log is thus placed, balks rolled up and put into position and leveled and the floor laid in the usual way.

261.—*For spans of 25 ft. or over*, when bottom can be touched clear across, some form of *trestle bridge* will be the easiest of construction.

262.—In Pl. 36, Fig. 1, the trestle consists of 6 legs, 4 vertical and 2 inclined; the two vertical legs on each side are fastened to two short sills by 2-in. pins. The ends of the two inclined legs are cut on an angle, driven into position and held by 2-in. pins passing through the transom from above. They serve as braces and supports. A short horizontal piece pinned to each pair of vertical legs supports the transom, which is also pinned. On top of the projecting ends of the vertical legs a cap piece can be pinned to form hand-rails. The trestle is made on shore, floated to its place in bridge, and erected with the aid of a float held at the proper distance from last trestle by a pole on each side, having the lengths of the spans marked by pins which engage the transoms of the trestles. The transoms are only temporarily lashed in position at first, but after the trestle is erected in its proper place the proper height for it is determined by bringing the 2 distance poles horizontal, or a little above the horizontal if a camber is to be given, then pinning or spiking on the short horizontal pieces. If accurate soundings have been made across the stream on the lines of the legs of the trestles, then the trestles can be completed on shore before launching. The balks are then run across and pinned and roadway finished. If there occurs unequal settling, the roadway can be raised by blocking up under the transoms on the short horizontal pieces.

263. Fig. 2 is another form of trestle called the Tie-block Trestle, consisting of only two legs, about 8 in. in diameter. The transoms are in pairs, across which two blocks are spiked at each end into notches, as shown. This trestle can be used on hard, uneven bottom. The trestle is formed on shore, held in shape by the rope, and rack stick across the top, then floated into place. Two poles, longer than two spans, are then run out; on the projecting ends are pins to prevent the trestle slipping off, and on near end a rope for fastening to transoms of second trestle back. Having caught the trestle on the ends of the poles under the transom, it can be raised to a vertical position by men bearing down on the rear, and held by means of ropes; it is then lowered into position, legs in a vertical plane. The transoms are then adjusted to their proper elevation by striking on under side, if too low, then tightening rope; or by slackening the rope and striking on upper side if too high. When properly adjusted, the rope is tightened, pinching the legs between the blocks; the braces are then spiked or pinned on, the rope removed, the balks laid and pinned, and the poles shoved out for the next trestle.

264.—Where sufficient lumber can be procured, the most expeditious and probably the best method will be as follows: (Fig. 3.) With the balks and chesses for each span, form a raft, or as many as may be desired, the length of a span. Form a trestle by placing 4 legs parallel and 4 ft. apart from center to center. Spike a pole across near the bottom and one near the top to keep them together. The first, or any trestle, having been set, float a raft against it and make fast; bring the trestle to be set up to the other end; force the legs under the raft a distance a little less than the depth of the water. Tie a rope around the outside legs at "f" with a bow-knot, to hold from slipping under, and others to the top pole, by means of which it is raised to a vertical position when it is dropped to the bottom by slackening off on lower ropes. As soon as it is dropped, another raft is brought up, tied, and another trestle put into position, and so continued. Each trestle, as soon as it is in position, is then capped by nailing two boards horizontally on opposite sides of legs with tops in same plane. (Fig. 3, a.) Braces are then spiked on the legs. Saw off the two inside posts even with the tops of boards. Spike a 2-in. plank across the top of posts and boards. Lay the balks, spike them, remove the raft, and move it into position to raise another trestle. If boards cannot be procured for capping, round sticks may be used as in Fig. 3, b, by cutting the two inside legs off 5 or 6 in. above the horizontal poles, then spiking two short pieces across the poles against the outside legs and one in center on which the cap piece or transom will rest. Advantages are—work can be commenced in any number of places at the same time; no accurate soundings required so long as poles are sufficiently long; capping and bracing do not retard work; different squads can be at work at same time, etc.

265.—If only axes and rope are available, trestles may be made by lashing their parts together. Fig. 4 shows the **Two-legged Trestle**. Having determined the height of the roadway above the bottom of the stream, mark this height from the butts on both legs, then mark the position of the transom on the legs, allowing for the thickness of the balks; also mark on the transom for the width of roadway between side-rails plus 3 ft., for points of crossing of legs, the distance apart of legs depending on width of roadway. Give the legs a splay outwards at the bottom of $\frac{1}{4}$ and mark on legs and ledger the points of lashing. All being ready, lay the transom on a couple of supports 3 or 4 in. high, inside the position of the legs, lay on the legs in their proper positions, on the legs lay the ledger. With the square lashing fasten the four points of crossing. Next, lay on the braces, butts and one tip on same side as ledger and one tip on side of transom. Lash the butts with square lashings. Square the trestle by making the diagonals equal, measuring from the centre of ledger lashing on one leg to the centre of transom lashing on opposite leg. When these diagonals are made equal the tips are lashed with square lashings and the braces at the middle with a cross lashing. Ledger and braces can be of rather light timber. The trestles can be floated into position and raised as already described, or run out and down from the end of the bridge, which is more difficult. They are kept vertical by lashing the balks to the transoms, and longitudinal bracing from one to another.

266. A Three-legged Trestle (Pl. 37, Fig. 1) may be made by first lashing two legs together considerably higher than the roadway is to be, then lashing the pry-pole just below to one of the legs, all with shear lashings. Stand the trestle up, spread out legs till butts rest on the vertices of an equilateral triangle whose sides are ⅓ height of trestle, then lash three light ledgers to the legs by round lashings. On the outside of the pry-pole and leg to which it is fastened are lashed short pieces, by square lashings, on which rest two longer pieces, separated by the legs, which are lashed together by the shear lashing. On these longer pieces rest the transoms. With these trestles lighter material can be used; they stand without bracing but are difficult to place; accommodate themselves to inequalities of surface; the roadway may be readily raised or lowered. If material is available, they are readily made with spikes.

267.—Fig. 2 is another form of trestle of four legs. Two two-legged trestles are made, one being 12 to 18 in. narrower than the other, depending on the size of legs, so that they will lock when put together. The transoms are placed on same side as ledgers, instead of on opposite sides. The butts of the single trestles are placed a distance apart equal to half the height, then locked at the top, the transoms lashed at the ends, longitudinal braces lashed at the ledgers, the tips tied and racked together. Sometimes used with light material, also as steadying points in a long bridge of two-legged trestles. One similar to it can be made of sawed timber and spikes and placed in position as shown in Figs. 3 and 4, if the materials are available.

268.—In sluggish streams with muddy bottoms and not over 6 ft. deep, where timber is abundant, crib piers may be used. (Fig.

PLATE 37.

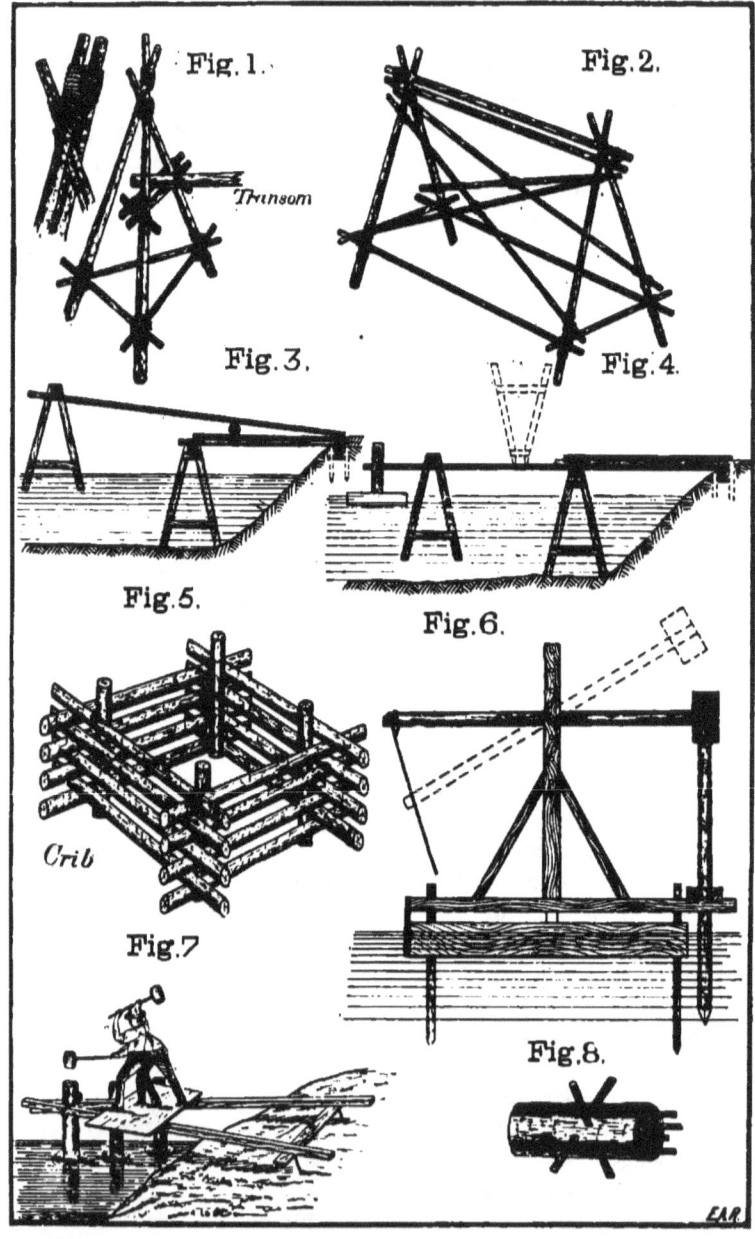

PLATE 38.

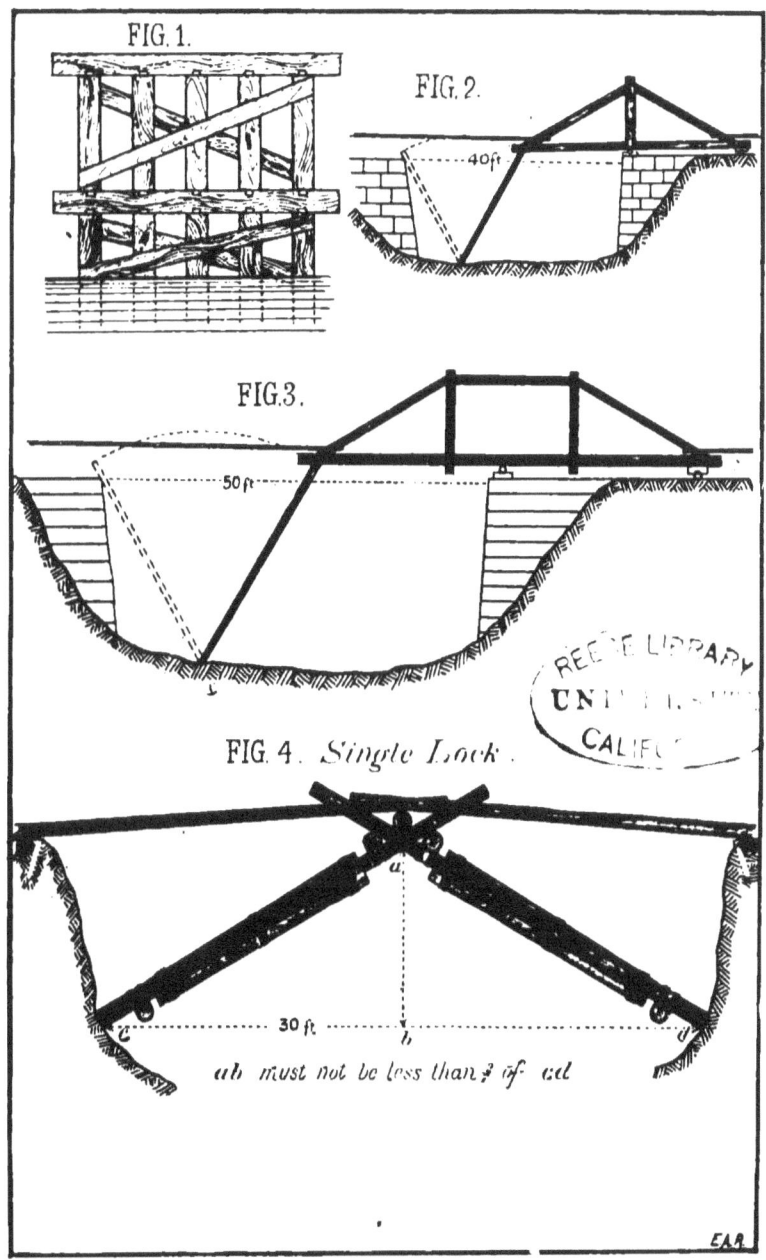

FIG. 1.

FIG. 2.

FIG. 3.

FIG. 4. *Single Lock*.

ab must not be less than ⅔ of cd

5.) The cribs are built in the woods, the foundation logs being pinned together, the others simply notched. The logs are then marked, taken down, carried or floated into position, and rebuilt, poles being generally set to mark the corners. As the crib is built up it gradually sinks, or a tray may be formed inside and loaded with stones. The balks and flooring are laid as usual.

269. *Pile bridges* are scarcely adapted to an emergency, from the time and preparation required in their construction, but on lines of communication, from the character of the bottom or the dangers from floating objects, resort may be had to them.

270. For driving the piles, a monkey (Fig. 8) is made of a block of wood 3 ft. long, 12 in. in diameter, with four 1.5 in. pins at top and four on the sides for handles. Four men standing on a platform on the pile drive it down, their own weight thus assisting, or they may be driven from a raft built as in Fig. 6. After the piles are driven they are straightened, braced, their tops sawed off level, the caps placed on and pinned (Pl. 38, Fig. 1), and the roadway laid as usual. Piles near shore may be driven as in Fig. 7.

271.—For crossings greater than 25 ft. and too deep to use any of the above forms, resort must be had to some form of *truss bridge*. The trusses may be put together either by lashing or with pins, or by combinations of both.

272.—Pl. 38, Fig. 2, represents the ordinary **King-post Truss** for spans up to 40 ft. The bridge is put together on the bank, then pushed forward half its length, using rollers under each truss, as shown. A trestle is then leaned forward from opposite bank, and, when truss is over it, the trestle is raised and the end of the truss carried over to the opposite bank.

273. Fig. 3 represents the **Queen-post Truss** for spans up to 50 ft. It is constructed and carried across similar to the preceding one.

274.— Fig. 4 shows the **Single-Lock** for spans of 30 ft. It consists of two frames similar to the two-legged trestle on Pl. 36, Fig. 4. A section of the gap is first marked out on the ground on each bank with the positions of the footings indicated. On these the legs are laid and the positions for lashing the transoms and ledgers marked. The frames are then put together opposite the position they are to occupy (one on each bank), butts towards the gap. One frame is made 15 to 18 in. wider than the other so they will lock, and the footings should be likewise prepared. The distance between legs at transom of narrower frame is 18 in. more than width of roadway between side-rails. With the above exceptions the frames are made like the two-legged trestles. The splay of the legs is very slight, generally about 1 ft. between transom and ledger. Stout stakes are then driven at the rear, fore and back guys are attached to the tips of each frame, the fore guys crossed over the stream, those of narrower frame in center. Foot ropes are also attached to each leg near the butts with timber hitches and a turn taken around the stakes at the rear. The frames are then shoved over the banks till they balance (Pl. 40, Fig. 1) then brought to a vertical position by hauling on the fore guys, and lowered into their places by easing off on the foot ropes, after which they are pulled over and locked. A couple of balks are then run out, then the fork transom is put into place and the balks rested on it. The remainder of the balks are then run out, placed on the fork transom, lashed, and the roadway completed as usual. If good places for footings cannot be secured, then other means must be provided.

275.— *For spans up to 45 or 50 ft.*, the **Double-Lock** (Pl. 39, Fig. 1) may be used. In this it will be noticed that the balk-bearing transoms are not the transoms first lashed to the frames in making them, but those which are sent out after the frames are in position. This must be remembered in marking the positions of the transoms on the legs of the frames. In this the two frames are made as described for *the single-lock*, except that they are of the same width. They are launched as described, and pulled forward until their tops are about ⅓ the span apart. Two straining beams are then run across, the road-bearing transoms fastened on top of them in the positions previously marked. The frames are held by the back guys until all is ready, when they are eased off and the bridge locked. The roadway is then laid as usual.

276.—*For spans greater than 45 or 50 ft.*, where timber of sufficient size is obtainable, the **Single Sling or Treble Sling** may be used. The frames are made as has been described, with the following additional observations:

In the *Single Sling* (Fig. 2), in marking the positions of the different spars, the three locking pieces must be at least 9 or 10 ft. above the roadway. The fork piece is hauled into position by snatch blocks lashed to the top of each leg of narrower frame,

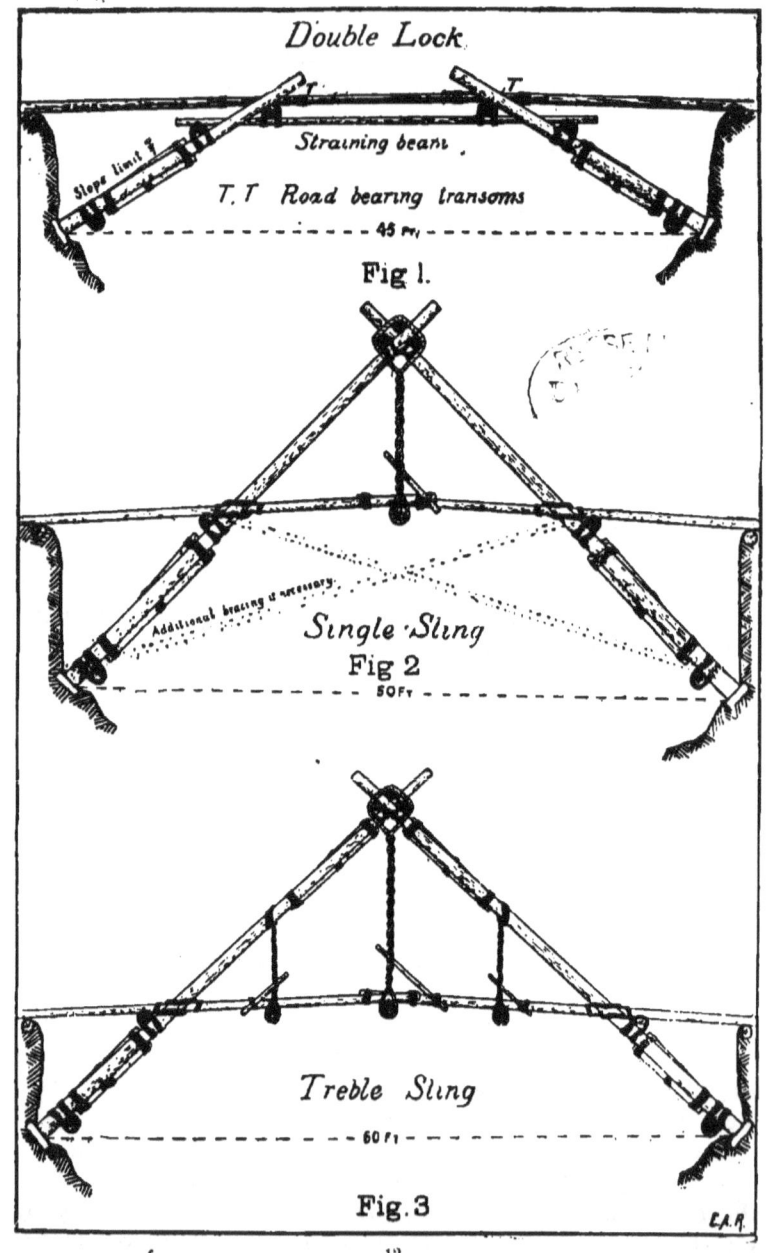

PLATE 40.

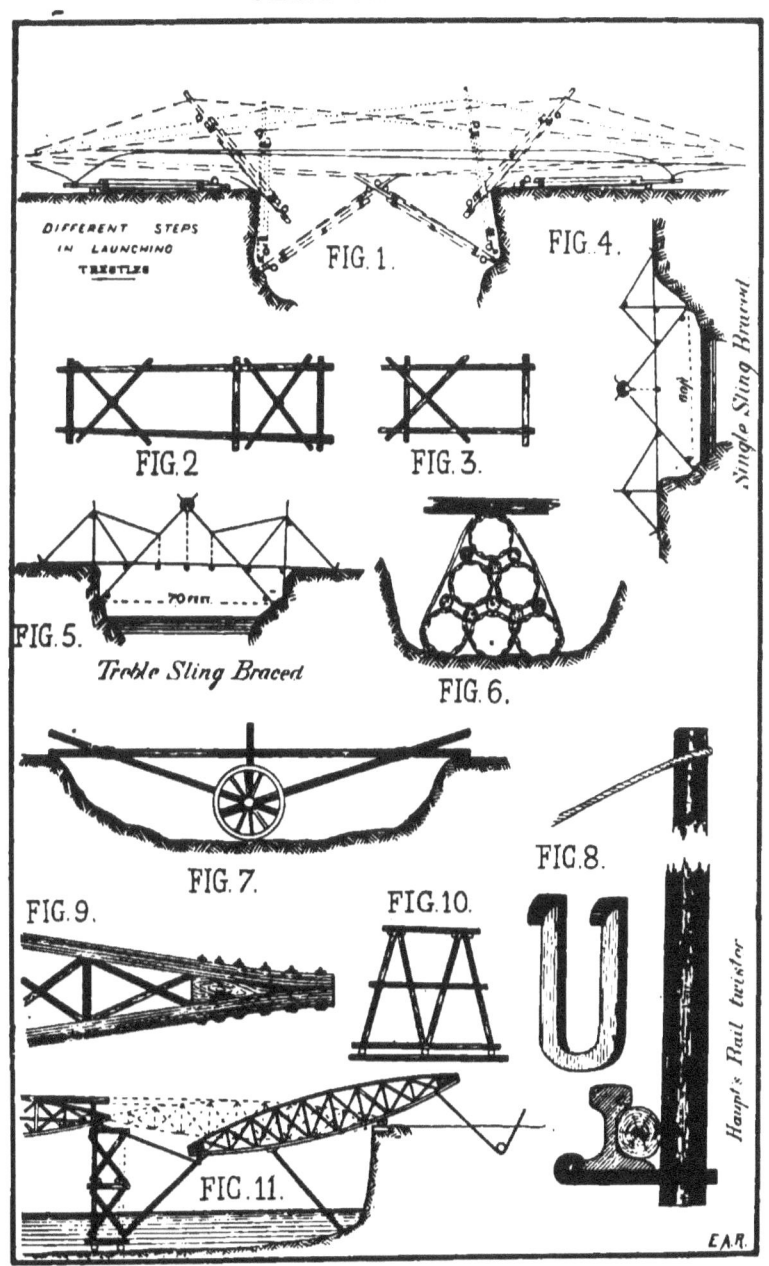

after which the blocks are used to get the center transom temporarily into position, when it is slung by the ropes that are to hold it, by taking several turns around it and the locking pieces without riding, and afterwards twisted up to the proper height with a pole.

277.—In the Treble Sling (Fig. 3) there are three slung transoms, one from the forks and one from the standards on each side of the middle. The frames are constructed as already described. (Pl. 40, Fig. 2, being one in plan.) If necessary, the frames may be strengthened by additional braces on them and further braced back to the banks by ropes attached to holdfasts and otherwise as suggested on Pl. 40, Figs. 4 and 5, vertical braces being shown in Fig. 3.

278.—Other expedients for crossing small gaps are the use of wagons in various ways for supports, brushwood made into gabions, fascines, etc. (Figs. 6 and 7.)

279.—A light, portable truss (Fig. 11) can be made, where boards are obtainable, by describing two arcs of circles with radii 151 ft., on opposite sides of a 60 ft. chord, then driving stakes on the arcs at intervals of about 2 ft., against which 5 layers on top and 6 layers on bottom of boards 1 in. thick x 12 in. wide, breaking joints, are bent and securely nailed together every 4 in. with tenpenny nails. The lower side of truss is made one board thicker than the upper and is completed by driving 6 in. spikes through every 6 in. This truss will be about 6 ft. deep, and, allowing 2 ft. at each end for resting on supports, will bridge a span 56 ft.

The sides are connected every 5 or 6 ft. by vertical pieces of plank and two 1-in. iron rods, the latter on the sides of the verticals, towards the middle. If iron rods are not obtainable, rope or wire should be wrapped around both and twisted tightly. The angles at the ends are filled with wedge-shaped pieces and the ends securely bolted, hooped, or wrapped. (Fig. 9.) For greater rigidity, light diagonal braces may be inserted in the panels. The top can be made straight instead of curved if so desired.

These trusses are used in pairs and are applicable to a variety of structures and to spaces of considerable width. Two such trusses with a central support of trestles, crib-work, or boats, may be used for 116 ft. (Fig. 11); three such trusses for 176 ft., etc. In experiments with such trusses in bridges, 1,800 lbs. per lineal foot has been applied before breaking; and by covering the boards with pitch and tar before nailing together, inserting ½ in. bolts in pairs every foot of length on lower side, and nailing boards against the edges, 3,500 lbs. per lineal foot was applied before breaking.

280. Suspension Bridges. For spans greater than 60 ft., and when timbers for frames cannot be procured, some form of suspension bridge might be used. Although applicable to longer spans, and the materials more easily transported, they take longer to make than other kinds.

The cables may be of iron chains, iron, steel or fibrous ropes, or of boards nailed together.

281.—Pl. 40a, Fig. 1, shows one with the roadway hung below the cables, with a camber $\frac{1}{30}$. At the center, the roadway should be at least 1 ft. below the cables. The width of roadway between side rails should be only slightly wider than wagon-wheel tracks. (Fig. 2.) On the banks, the cables are supported by timber piers (Fig. 3), having a broad cap (Fig. 4), rounded on top, over which they pass at a distance apart of 9.5 ft. The cables must be securely anchored at the rear to heavy logs sunk 4 or 5 ft. in the ground, or otherwise, and drawn in until the sag is only $\frac{1}{10}$ or $\frac{1}{12}$ of the span.

282. In Fig. 5, part of the roadway is hung below and a part rests on the cables, the greatest slope of road being 1 on 6 for 100 ft. span and $\frac{1}{10}$ sag. The cables are only 7 ft. apart.

283.—In Fig. 6, the roadway is built on trestles supported on the cables. For spans 130 ft., sag $\frac{1}{12}$, the frames form the sides of equilateral triangles of 10 ft. each. To construct it, the curve of the cables is traced on the ground, the trestle legs laid on it and marked where they cross the road and cable; those for each half of the bridge are ranged in order on the banks, connected together as placed on the cables and hauled out, connected at the center, the curve of the cables adjusted and the bridge completed.

284. Fig. 7 is a suspension bridge of which the cables are made of boards nailed together in several thicknesses laid horizontally, breaking joints; the ends are spread apart and wedge-shaped blocks inserted and anchored by several rows of posts, as shown in Fig. 8. Each cable, as made, is drawn across by ropes, anchored, and the trestles placed from both ends at the same time. Last of all, spikes long enough to reach entirely through the cable are driven every 4 to 6 in.

285.—Fig. 11 is a similar bridge supported on trestles 16 ft. long, not exceeding 20 ft. high, placed at intervals of 40 ft., over which suspends the two board cables, 14 ft. apart. On these are placed low trestles, 3 ft. high, dividing the spans into lengths of 20 ft. each; 25 ft. balks are used and the roadway laid as usual. The cables are made of six thicknesses, of 1 in. boards 12 in. wide, breaking joint, nailed and spiked every 4 to 6 in., and bolted by pairs of $\frac{1}{2}$ in. bolts every foot. Three thicknesses of boards are first nailed together and drawn across, the ends anchored, and then the other three boards added.

286. For light foot bridges (Fig. 10), across narrow gaps, wire from fences, if available, could be used for the cables by twisting a number together and passing them over crotches of trees and anchoring to stumps, etc., in rear, and then laying the walk similar to some of the methods previously shown.

287.—So various are the conditions to be met in constructing bridges that seldom will any one type meet the requirements, but by the application of good judgment and resource, with the suggestions here offered, almost any gap of reasonable width may be crossed, if not by one type or another, then by a combination of several to meet the emergency.

The varying strength of timbers makes it almost impossible to give exact dimensions for the different spars to be used in the different types, but a general idea may be obtained below of the amounts and average dimensions of medium strength timber, as yellow pine. For weaker timbers some of the sizes will have to be increased, while for stronger ones there will be an excess of strength if the sizes given are adhered to, but the desire to be on the side of safety warrants the use of amounts which might, by a careful mathematical calculation, appear to be excessive.

The timbers for transoms, ledgers, braces, balks, flooring and side rails should be selected of as nearly a uniform diameter throughout as possible and will be so considered in giving dimensions. For legs or standards the diameter at tip will be given.

For a 9 ft. roadway with 15 ft. spans, 5 balks 20 ft. long x about 6 in. in diam. are used, and placed $2\frac{1}{4}$ ft. apart from center to center. For the flooring are used poles 11 to 12 ft. long x 4 to 5 in. in diam. For side rails 2 poles 20 ft. long, 4 to 6 in. in diam.

For each Six-legged Trestle (Pl. 36, Fig. 1) 4 vertical and 2 bracing legs 6 in. diam., 1 transom 12 ft. x 8 in., 2 foot pieces 3 ft. x 8 in., 10 oak pins 2 in. diam.

For each Tie-block Trestle (Fig. 2) 2 legs 8 in. diam., 2 transoms 15 ft. x 8 in., 4 tie blocks 2 ft. x 5 in. x 6 in., 2 braces 3 ft. x 2 in. x 6 in., 24 spikes, 1 rope, 1 rackstick.

For each Capped Trestle (Fig. 3) 4 legs 8 in. diam.; 2 braces 12 ft. x 4 in.; 2 braces 15 ft x 5 in.; 3 boards 12 ft. x 2 in. x 12 in.; 4 ropes, spikes.

For each Two-legged Trestle, lashed, (Fig. 4) 2 legs 4 ft. longer than height of trestle, 5 to 7 in. tip; 1 transom 15 ft. x 9 in.; 1 ledger 16 ft. x 4 to 6 in.; 2 braces 3 to 5 in. diam.; 6 ropes 30 ft. x ¼ in. diam.; 3 ropes 15 ft. x ¼ in. diam.

For each Three-legged Trestle, lashed, (Pl. 37, Fig. 1) 6 legs 3 to 5 in. tip; 4 transom bearers 6 ft. x 3 to 4 in.; 4 sticks 2 ft. x 2 to 3 in.; 6 ledgers 2 to 3 in. diam.; 1 transom 15 ft. x 9 in.; 12 ropes 30 ft. x ¼ in. diam.; 6 ropes 15 ft. x ¼ in. diam.

For each Four-legged Trestle, lashed, (Fig. 2) twice the amount given for each two-legged trestle, plus 2 ledgers, and 6 lashings 15 ft. long.

For each Single Lock (Pl. 38, Fig. 4) 4 legs 22 to 25 ft. x 7 in. tip; 1 ft. transom 15 ft. x 10 in.; 2 frame transoms 15 ft. x 6 in.; 2 ledgers 15 ft. x 4 to 6 in.; 4 braces 20 ft. x 3 in.; 2 shore sills 15 ft. x 6 in. *Lashings*, 4 transom 50 ft. x ¼ in.; 10 ledger and brace 30 ft x ¼ in.; 10 balk 20 ft. x ½ in.; 4 foot 50 ft. x 1 in.; 8 guy 150 ft. x 1 in.

For each Double Lock (Pl. 39, Fig. 1) 4 legs 22 to 25 ft. x 7 in. tip; 2 straining beams 25 ft. x 8 in.; 2 road transoms 15 ft. x 10 in.; 2 frame transoms 15 ft. x 6 in.; 2 ledgers 15 ft. x 5 to 6 in.; 4 braces 20 ft. x 3 in.; 2 shore sills 15 ft. x 6 in.; *Lashings*, 8 transom 50 ft. x ¼ in.; 14 ledger and brace 30 ft. x ¼ in.; 10 balk 20 ft. x ¼ in.; 4 foot 50 ft. x 1 in.; 8 guy 150 ft. x 1 in.; besides axes and other tools, and anchorages, holdfasts, etc., on banks.

For each Single Sling (Fig. 2) 4 legs 35 to 45 ft. x 6 in. tip; 3 top and fork transoms 15 ft. x 6 in.; 3 road transoms 15 ft. x 10 in.; 2 ledgers 15 ft. x 4 to 6 in.; 4 braces 20 ft x 3 in.; 2 shore sills 15 ft. x 6 in.; 10 balks 30 ft. x 6 in.; 4 side rails 30 ft. x 4 to 6 in. Lashings of same number, size and length as for Double Lock. Stiffening will require additional spars and lashings, depending upon the method used.

PLATE 40a.

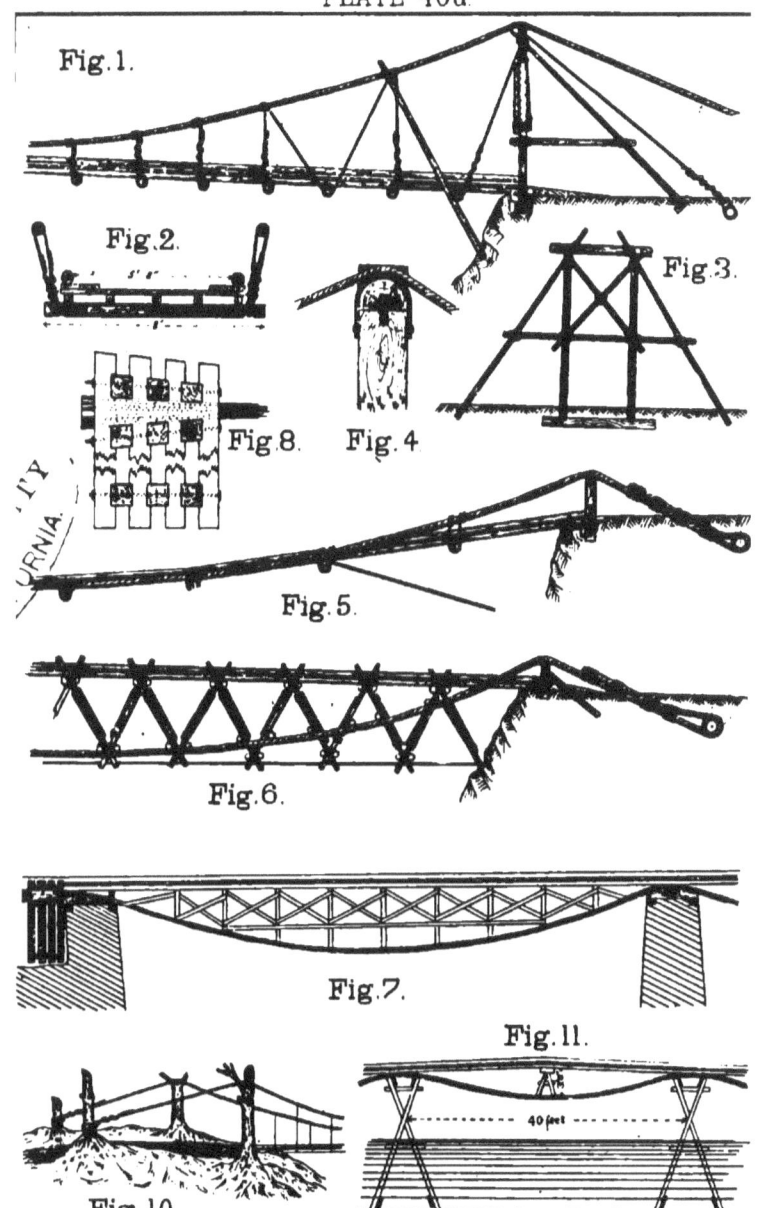

SPAR BRIDGES. 175

For each Treble Sling (Fig. 3) 4 legs 50 ft. x 6 in. tip; 5 road transoms 15 ft. x 10 in.; 3 top and fork transoms 15 ft. x 6 in.; 2 lower ledgers 15 ft. x 4 to 6 in.; 4 lower braces 20 ft. x 3 in.; 4 upper braces 18 ft. x 3 in.; 2 shore sills 15 ft. x 6 in.; 15 balks 5 ft. longer than $\frac{1}{3}$ span x 6 in.; 6 side rails 5 ft. longer than $\frac{1}{4}$ span x 4 to 6 in.; 6 sling racking sticks 10 ft. x 4 in. *Lashings*, 4 foot 50 ft. x 1 in.; 8 guy 150 ft. x 1 in.; 24 ledger and brace 30 ft. x $\frac{1}{4}$ in.; 8 transom 50 ft. x $\frac{1}{2}$ in.; 40 or 50 balk 20 ft. x $\frac{1}{3}$ in. Stiffening will require additional spars and lashings, depending upon the method used.

For Suspension Bridge 200 ft. long (Pl. 40a, Fig. 1) 4 to 8 cables 180 ft. x 1 in.; 16 cable seizing of yarn 18 ft. long; 12 lashings 50 ft. x $\frac{3}{4}$ in.; 10 lashings 30 ft. x $\frac{1}{4}$ in.; 100 lashings 20 ft. x $\frac{1}{5}$ in.; 2 steel wire cables 400 ft. x 1$\frac{3}{8}$ in.; 4 standards 26 ft. x 10 in. tip; 4 braces 22 ft. x 3$\frac{1}{4}$ in. tip; 2 caps 12 ft. x 10 in.; 2 sills 15 ft. x 10 in.; 4 back struts 36 ft. x 4 in. tip; 4 side struts 32 ft. x 3 in. tip; 4 cable props 30 ft. x 5 in. tip; 4 horizontal ties 30 ft. x 3 in. tip; 21 transoms 10 ft. x 6 in.; 80 balks 13 ft. x 6 in.; 40 side rails 20 ft. x 6 in.; for anchorages 16 spars 5 ft. x 7 in. tip; 2 spars 16 ft. x 20 in.; 2 spars 16 ft. x 12 in. $\frac{3}{4}$ round; 10 spars 16 ft. x 8 in. $\frac{3}{4}$ round; 4 back ties 50 ft. x $\frac{3}{8}$ in. steel rope; 4 ties 35 ft. x $\frac{1}{4}$ in. steel rope; 40 slings total 600 ft. x $\frac{1}{4}$ in. steel rope; 4 guys 50 ft. x 1 in.; 4 rope ladders.

For Suspension Bridge 100 ft. long (Fig. 5) 4 to 8 cables, 12 cable seizings, 4 lashings, 12 lashings, 30 lashings as above; 2 cables 180 ft. x 3 in. hemp or 2 in. steel; 2 anchor spars 18 ft. x 15 in.; 10 transoms 12 ft. x 4 in.; 4 balks 25 ft. x 6 in.; 10 side rails 20 ft. x 4 in; materials for piers depending on circumstances.

For Suspension Bridge 130 ft. long (Fig. 6) 4 to 8 cables; 12 cable seizings, 9 lashings, 104 lashings, 280 lashings as above; 4 cables 200 ft. x 2$\frac{3}{8}$ in. hemp or 1$\frac{1}{4}$ in. steel; 2 anchor spars 18 ft. x 18 in.; 44 trestle legs 13 ft. x 3 in. tip; 44 braces 15 ft. x 2 in. tip; 22 transoms 9 ft. x 4 in.; 80 ledgers 12 ft. x 2 in.; 20 cable ledgers 12 ft. x 5 in.; 2 shore sills 10 ft. x 5 in.; 48 balks 14 ft. x 5 in.; 28 side rails 20 ft. x 5 in.; materials for piers depending on circumstances.

Besides the above materials, there will be required tools for cutting timber, tackles for raising frames, shovels, pickets, etc., and, where not mentioned, the ordinary amounts of balks, chess, side rails, etc.

CHAPTER XVI.—Floating Bridges.

288.—The passage of a stream may be effected, in many cases, as described in the preceding chapter. If the methods there laid down are not suitable or expedient, and the stream cannot be forded, then resort must be had to ferrying by boats, rafts, flying bridges, or to floating bridges.

289.—The selection of a place and means of crossing a river is determined by a reconnaissance, which should be as detailed and extensive as circumstances will permit, and embrace the following:—
(a) The nature of the banks.
(b) The nature of the bed.
(c) Position and depth of fords.
(d) Strength of the current.
(e) Whether tidal or otherwise.
(f) Probability and extent of floods.

290. Fords. A stream with a moderate current may be forded by infantry when its depth does not exceed 3 ft., and by cavalry and carriages when its depth is about 4 ft. The requisites of a good ford are:—
(a) Banks low, but not marshy.
(b) Water attaining its depth gradually.
(c) Current moderate.
(d) Stream not subject to freshets.
(e) Bottom even, hard, and tenacious.

291.—In a mountainous country, the bed of a stream is likely to be covered with large stones, rendering the passage of carriages impracticable. In level countries, the bed of the stream may be composed of mud or quicksand, rendering passage by fording impossible. In some cases, the bottom is composed of fine sand, which is hard enough, but which, by the action of the hoofs of the

PLATE 41.

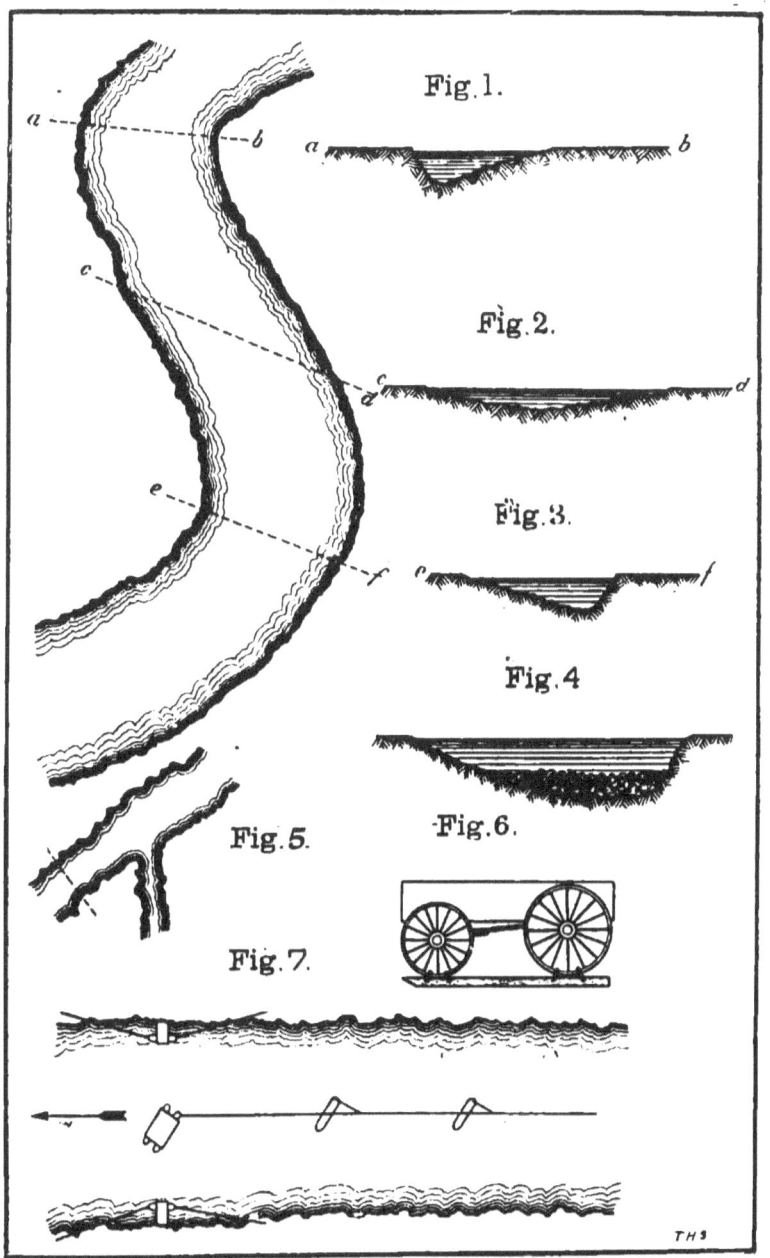

animals, is stirred up; the current then carries the sand away and the ford is deepened, perhaps so much as to become unfordable. The best bottom is coarse gravel.

292.—Fords are usually found in the wider and more rapid parts of a stream. A straight reach gives the most uniform depth. At bends, the depth will generally be greater at the concave bank and less at the convex. (Pl. 41, Figs. 1 and 3.)

293.—To determine the position of a ford :—

(1) A number of mounted men may be sent across wherever there is a probability of the river being shallow enough.

(2) *Most certain method.* Float down the stream in a boat, keeping in the swiftest part of the current, where the water is usually deepest. Hang a sounding line of the proper length over the stern. When this touches bottom, sound across the stream.

When a ford is discovered, it should be marked by stakes; remarkable objects on the shore should be noted; and a stake planted at the water's edge and marked, in order that any rise in the water may be at once evident.

294.—A stream, otherwise unfordable, may be passed :—

(1) By crossing it in a slanting direction. (Fig. 2.)

(2) When the unfordable portion is not over 8 or 10 yards, this may be filled in with fascines loaded with stones. (Fig. 4.)

(3) When the bottom is muddy, it may be covered with bundles of coarse grass, rushes, or twigs, sunk by means of stones.

(4) A portion of the water may be diverted from its natural channel. (Fig. 5.)

295.—In passing a stream by fording, if it is deep and the current at all swift, the following precautions should be taken :—

(a) Troops passing in column should do so at a considerable interval, in order to avoid choking the stream.

(b) If boats are to be had, a few should be stationed below the ford, to assist men who may be carried down by the current.

(c) If boats cannot be procured, mounted men may perform the duties described in the foregoing provision.

(d) In place of provisions "b" and "c," a life line, held up by casks, may be stretched across the stream.

(e) In order to break the force of the current, cavalry may be stationed in the stream, above the point of crossing.

296.—After a freshet, a ford should always be re-examined, lest some alteration may have taken place in the bed of the stream. The banks of a stream to be forded should, if necessary, be scarped.

The velocity of a stream may be determined by throwing in a light rod, so weighted as to stand vertically. Note the distance passed over in a certain number of seconds; then, $\frac{7}{10}$ the mean number of feet per second gives the velocity in miles per hour.

297.—Ice. In high latitudes, during the winter, rivers are frequently covered with ice of sufficient thickness to sustain the heaviest loads. This means of passing a stream should be used with great circumspection. A change of temperature may not only suddenly destroy the natural bridge, but render the river impassable by any method, for a considerable time, in consequence of floating ice.

298.—Ice, in order to allow of passage, should be of the following thickness:

For Infantry, single file, 2 yds. distance, on a line of planks..	2 in.
For Cavalry or light guns, with intervals.............	4 in.
Heavy field-pieces..................................	5 to 7 in.
Heaviest loads.....................................	10 in.

299.—When there is any doubt as to the strength of the ice, two tracks of plank may be laid for the carriage wheels to run on, or the wagon may be transformed into a kind of sled by fastening two planks under the wheels. (Fig. 6.)

The thickness of ice may be increased, when the temperature is low, by throwing water on it. When a stream is frozen on each side but open in the middle, in consequence of the velocity of the current, a boom stretched across the open space will often check the velocity sufficiently to allow the water to freeze.

300.—If a stream cannot be forded, it may be crossed by ferrying or by constructing a bridge. Ferrying may be by boat, raft, or flying bridge; rowed, sheared, or hauled across.

301. Ferrying by Boat. All boats available should be collected and taken to the chosen point of passage. The banks of the stream, if steep, should be scarped to facilitate embarkation. The landing should be farther down the stream than the point of starting. The boats should be arranged along the shore and numbered. Entrance to the boats should be by file, the soldiers taking positions on opposite sides alternately. Where the water is shallow near the shore, the boat should not approach the bank so closely as to ground as the men file in. The unloading should be made in the same manner as the embarkation, *i. e.*, by file alternately from each side of the boat. During the transit, the men should remain in position and not rise up suddenly when the boat lurches.

In passing artillery, the piece should be dismounted. Horses should, ordinarily, be made to swim. However, if the boats are large enough, the bottoms may be covered with plank, and the horses placed crosswise, facing alternately up and down stream.

302.—Ferrying by Raft. Rafts may be made of logs, lumber, casks, and other material suitable for the purpose. Their construction is the same as explained for piers of bridges, hence only two expedients will be mentioned here.

303.—The Canvas Raft. No other material being available, small rafts can be constructed by the use of canvas about 8 x 12 ft., and brushwood. Wet the canvas to make it water-proof, and lay it out on the ground. Across the width place sticks in layers, the longest near the middle. The sides should be strengthened by heavy sticks placed lengthwise. The pile of sticks should be about 4 ft. wide in the center and sloping off slightly towards the ends, 3 ft. high and 8 ft. long. Over this pile a second piece of canvas, after being wet, should be placed. The sides of the canvas on the ground are now drawn over toward each other and lashed securely with a lariat. The ends are folded neatly, brought up towards each other, and lashed. If care is taken to wet the canvas thoroughly and make it water-tight, this raft will carry three troopers with their arms and accoutrements. By lashing several together, a larger number of men, with their arms and accoutrements, can be carried.

304.—Rafts of Skins. Bags, made of the skins of animals, inflated with air or stuffed with hay or straw, can be utilized for crossing streams, and have been used from ancient times.

305.—Rafts are more suitable for the embarkation and landing of troops of all arms than boats. They will carry a larger number each trip, are not so easily injured by the fire of the enemy, and draw little water. On the other hand, they cannot be navigated with the same facility as boats, move much more slowly, and hence keep the troops much longer under fire; cannot be directed with certainty on a fixed point when the stream is rapid, and, if the passage is to be effected secretly, the time required for their construction is too long to admit of their use.

306.—The Floating Bridge. This may be formed of two boats covered with a platform, constructed as follows:—(Pl. 42, Fig. 1. The lashings and side rails are omitted.) From 5 to 7 beams of the same thickness are laid across the two boats, the intervals between the beams being equal, and such that the covering planks extend 1 ft. beyond the extreme beams. The interval between the boats is such as to allow the beams to extend 2 ft. beyond the gunwales. The beams are lashed to the boats, the covering planks are kept in place by 2 side rails, laid directly over the outer beams, and lashed down to them; the extreme planks should be nailed down.

The floating bridge can be navigated by oars with nearly the same facility as a boat.

307.—The Rope Ferry. The rope ferry, which is used in sluggish streams, consists of a floating support, either a raft, floating bridge, or a large boat. It is drawn by hand along a rope stretched from shore to shore.

308.—The Trail Bridge. This is employed in streams not more than 150 yds. in width, and whose current is not less than 3 ft. per second, or $2_{1\over 5}$ miles per hour. The rope must be maintained above the surface of the water, and, consequently, must be drawn very tightly by means of a windlass, blocks, and falls, or similar expedients; it must, also, at each bank, be raised some distance above the water. (Pl. 42, Fig. 3.)

The raft, or boat, is attached to a pulley, which runs on a sheer line, and by means of a rudder is given such a position that its side makes an angle of about 55° with the direction of the current. The angle of 55° with the current divides its force against the side of the boat into two components: one, perpendicular to the sheer line, which is counteracted by the resistance of this line; the other, parallel to it, which moves the boat. A boat for this kind of ferry should be narrow and deep, with nearly vertical sides.

If a raft is used, it should be lozenge-shaped, the acute angle being about 55°. When two sides are parallel to the current, the

PLATE 42.

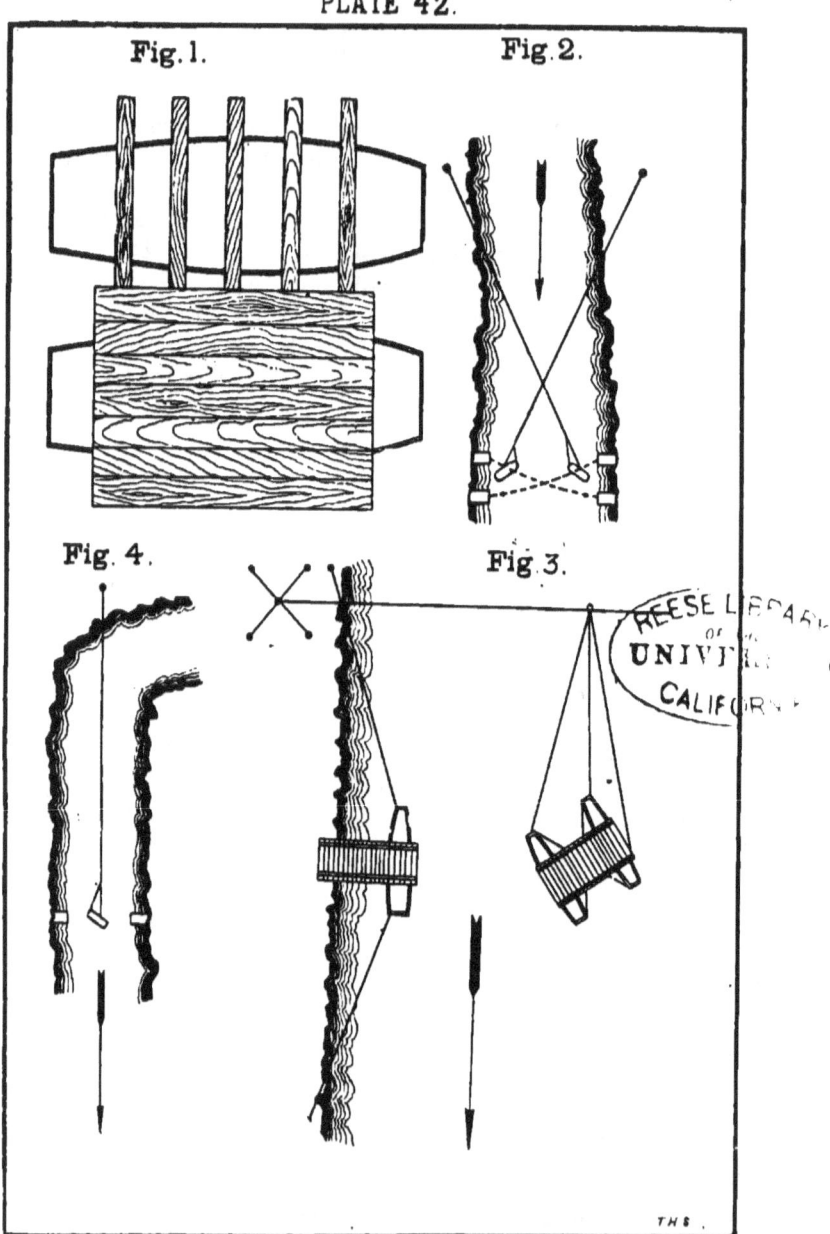

PLATE 43.

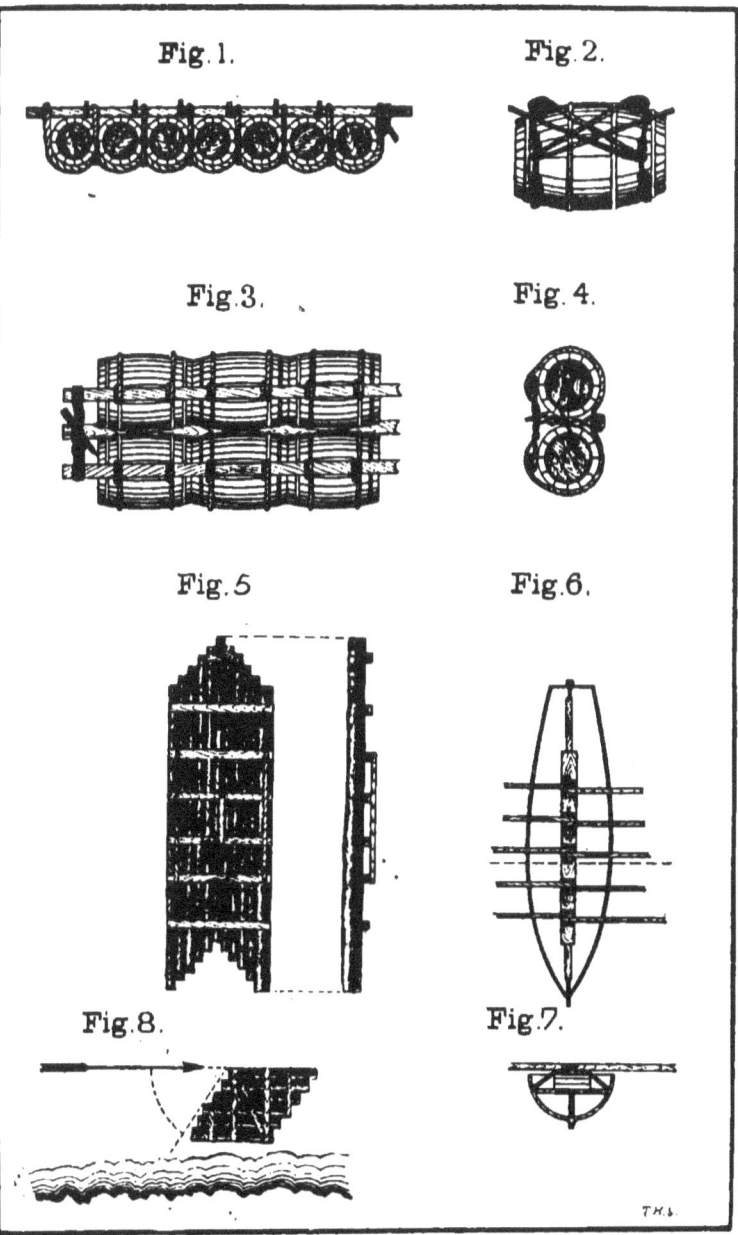

up-stream side will then be in the most favorable position for passage. (Pl. 43, Fig. 8.)

309.—The Flying Bridge. The character of the float for this ferry is the same as in the preceding case. (Pl. 41, Fig. 7; Pl. 42, Fig. 2.) This bridge is resorted to when the stream is wider than 150 yds. The strain on the sheer line being very great, it is replaced by a cable anchored in mid-stream, in which case the float would swing between two landing piers; or by two cables, one anchored on either bank, the float swinging between four piers. The latter requires less skill in manipulation. The angle which the float makes with the current is the same as that of the "trail" bridge. A sharp bend may be utilized for anchoring the cable, as shown in Pl. 42, Fig. 4.

The length of a swinging cable should be $1\frac{1}{2}$ to 2 times the width of the stream. The cable should be supported on intermediate buoys or floats, to prevent it dragging in the water.

310. Floating Bridges are composed of a roadway and its supports. The roadway is explained in the preceding chapter. The supports are floating, as pontoons, boats of commerce, rafts of barrels, logs, lumber, inflated skins of animals, or other material. The supports are called *floating piers*. It is from the character of the support that the bridge derives its name.

311.—In constructing a floating bridge, the *site* should be first selected and the *width* of the stream measured.

In selecting a *site*, the following points should be noted:—

(a) Proximity to a road. As the approaches to floating bridges, having frequently to be constructed across meadows, give much trouble, they should be as short as possible. For a similar reason, marshy banks are undesirable.

(b) The bed of the stream, if anchors are required, should afford good holding ground.

(c) A bridge can be best defended if constructed at a re-entering bend of a river.

(d) Use can frequently be made of islands to economize material.

312.—In measuring the *width* of the stream, if it cannot be done directly, some one of the methods explained in Chap. III. can be used.

313.—It should be remembered that a wide roadway gives greater steadiness than a narrow one. In making calculations for

188   FLOATING BRIDGES.

buoyancy, the weight of a 9 ft. roadway may be taken at 80 lbs. per running foot.

314.—Piers. Of whatever material the floating pier is made, the following points should be observed:—

(1) The available buoyancy of each pier should be sufficient to support the heaviest load that can be brought on one bay of the bridge.

(2) Piers should be connected with each other, at their extremities, by tie balks or lashings.

(3) To insure steadiness, the length of a pier should be at least twice the width of the roadway.

(4) The water way between piers should, if possible, be more than the width of the piers, never less.

315. Piers of open boats. In forming a pier of open boats, the following precautions should be taken:—

(1) The boat should not be immersed deeper than within 1 ft. of the gunwale.

(2) If the water is rough, or the current extremely swift, a boat should not be immersed deeper than within 1 ft. 4 in. of the gunwale.

(3) Boats should be placed *in bridge* with bows up stream or toward the current.

(4) If the stream is *tidal*, the bows of the boats should be alternately up and down stream.

(5) Unless the boat is very heavy and strong, the balks should not rest on the gunwales; a central transom should be improvised by resting a timber on the thwarts or seats, blocking up from underneath and bringing the weight directly on the keelson. (Pl. 43, Figs. 6 and 7.)

(6) Large boats should be placed where the current is swiftest, also as the first and last boats in bridge.

316.—The buoyancy of a boat may be found by one of the following rules:—

(1) To find the *available* buoyancy load the boat with unarmed men to a safe depth. Multiply the number of men thus loaded by 160. The result will be the *available* buoyancy in *pounds*.

(2) If the boat is afloat and empty, the *available* buoyancy may be found by calculating the volume between the then water line and the "safe load" line, and multiplying by $62\frac{1}{4}$.

(3) To find the *total* buoyancy. If the boat is of nearly uniform section, the area of the section multiplied by the length of the boat will give the cubic contents. A cubic foot of water weighs 62½ pounds.

Hence, if the dimensions of a boat are taken in feet, the contents will be cubic feet, and this, multiplied by 62½, will give the *displacement* of the boat; from this subtract the weight of the boat; this will give the *total* buoyancy.

317.—To find the length of a bay.—First find the available buoyancy of the boat. Then find the weight per running foot of the load the bridge is to bear, and to this, add the weight per running foot of the roadway. Divide the available buoyancy by this sum. The quotient will be the distance from center to center that boats should be placed apart. Thus:—Suppose the weight per running foot is 480 lbs., that the roadway is 80 lbs. per running foot. ∴ 480+80=560. The available buoyancy is found by one of the preceding rules, to be 5,600 lbs. ∴ 5,600÷560=10.

318.—The open boats may be:—(1) Those of commerce usually found on streams. (2) Regularly constructed pontons. (3) Improvised boats.

The first class requires no description. The second class comprises the canvas ponton used in the Advance Guard Train, and the boat or barge used in the Reserve Train, of the U. S.

319.—The table below gives the dimensions of the ponton in the U. S. Advance Guard Train, shown in Pl. 47.

Canvas Ponton 21' x 5' 4"x2' 4". Weight, 510 lbs.
Balks 22'x4¼"x4¼".
Side Rails same as Balks.
Chess 11'x12"x1¼".

WEIGHTS FOR ADVANCE GUARD TRAIN.

	Wagon.	Load.	Total.
	lbs.	lbs.	lbs.
Ponton	1,750	1,985	3,735
Chess	1,750	1,856	3,606
Trestle	1,750	2,060	3,810
Tool	1,700	1,938	3,638
Forge	1,217	1,166	3,383

320.—The table below gives the dimensions of the Ponton in the U. S. Reserve Train, shown in Pl. 48.

Ponton 31'x5' 8"x2' 7". Weight 1,600 lbs.
Balks 27'x5"x5" for a 20' span.
Trestle Balks 21'8"x5"x5".
Chess 13'x12"x1¼".
Side Rails same as Balks.

WEIGHTS FOR RESERVE TRAIN.

	Wagon.	Load.	Total.
	lbs.	lbs.	lbs.
Ponton	2,200	2,900	5,100
Chess	1,750	2,280	4,030
Trestle	2,200	2,635	4,835
Tool	1,700	2,100	3,800
Forge	2,217	1,166	3,383

321.—**Improvised Boats.** To reduce the amount of transportation required by an army is a very important consideration, hence the value of the following expedients.

322.—**The Crib Ponton.** This boat is 18 ft. long, 5 ft. wide, 2¼ ft. deep and covered with canvas. Construction. (1) Let stakes 4 ft. long, 2¼ in. in diameter, and 2 ft. apart, be driven into the ground, (Pl. 44, Figs. 1, 2 and 3), to the depth of about 1 ft., so as to enclose a space of the proper size for the top of the boat. The tops of the stakes should be in the same horizontal plane. This may be tested by placing a straight edge on them. Those that are too high can then be driven down.

(2) Nail boards against the outside of the stakes, extending 4 in. over their tops.

(3) Cross-pieces, of the same diameter as the stakes, are laid across the tops and pinned down upon them with wooden pins.

(4) Nail the side boards to the ends of the cross-pieces, and cover the bottom of the boat, which in its inverted position is now on top, with boards, and nail the projecting edges of the sideboards to the bottom securely.

(5) Finish boarding sides and ends to the proper depth.

(6) The frame is now ready to be covered with canvas. For a boat of the foregoing dimensions, the canvas should be 23¼ ft. x

PLATE 44.

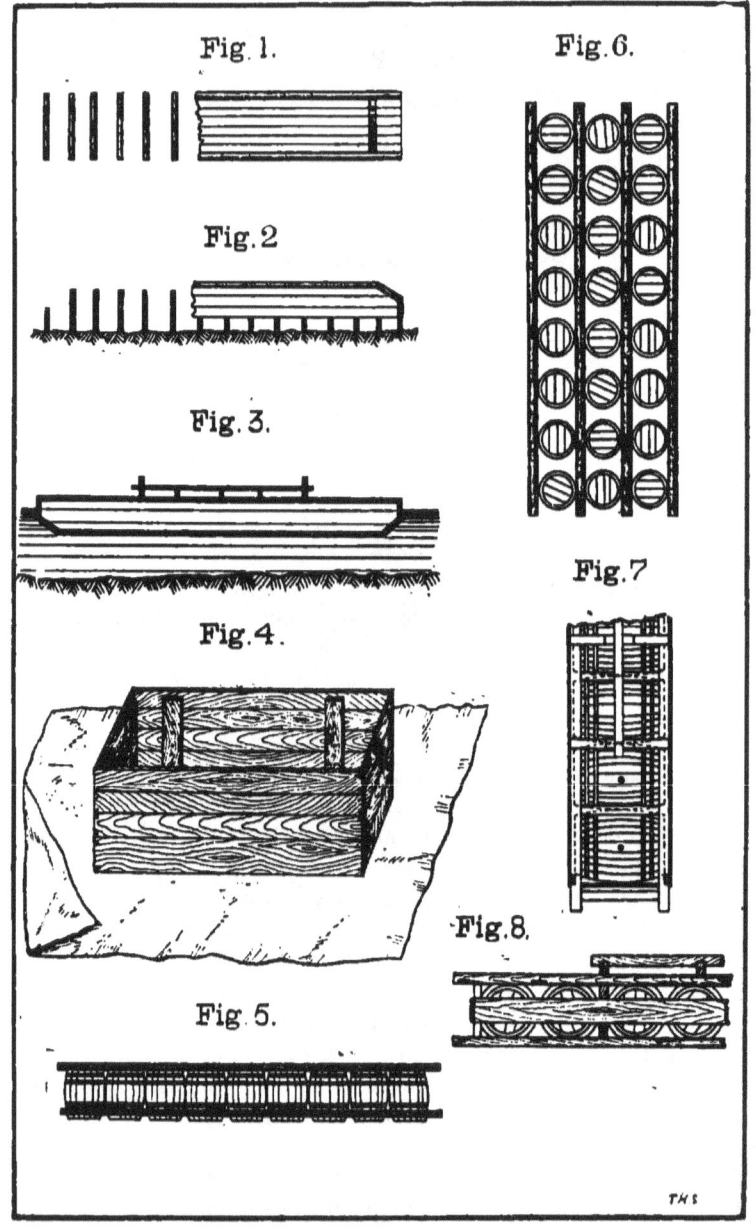

10¼ ft., about 6 in. being allowed for lap. The canvas may be put together in any number of pieces by daubing the edges of the seams with a waterproof composition and connecting them with ordinary carpet tacks.

(7) The canvas having been prepared, it should now be coated with a waterproof composition. Tallow, put on hot, will do if nothing better can be found.

(8) Place the canvas on the frame, coated side downward. Tack the canvas to the frame and cover with waterproof composition.

(9) Spike or pin 2 or 3 stout poles to the bottom longitudinally (not shown in drawing) to keep the bottom from abrading. If these poles are allowed to project about 6 in. at each end, they will assist in launching.

(10) Loosen the stakes from the ground by means of levers. Turn the boat over and saw off the stakes about 2 in. below the top edge of the side and end boards.

(11) Pin stout poles to the top of the stakes on the sides and ends, and nail the side and end boards securely to them.

(12) The side poles should project about 6 in. beyond the ends corresponding to those on the bottom, and be lashed to the bottom poles by means of a rope loop and rack stick. (Not shown in drawing.)

(13) Turn the canvas over the top poles and tack it down. The boat is finished.

323.—The Box Ponton. In localities where planks and boards can be conveniently procured, pontons may be constructed very expeditiously by placing two partitions of 2 in. plank, each 5 ft. long and 2½ ft. high, in parallel positions, on the top and ends of which boards are nailed. (Pl. 44, Fig. 4.) The box thus formed to be covered with pitched canvas, as described in the mode of constructing crib pontons. Where sound lumber is at hand, the box ponton will be more easily and expeditiously constructed than the crib ponton, but if plank is not at hand it may be preferable to use poles or split timber rather than wait for it.

324.—Wagon Body Ponton. Ordinary wagon bodies, covered with waterproof canvas or india rubber blankets, may be used either as boats or pontons. The small capacity of the wagon body requires such pontons to be placed more closely, to compensate for it.

325.—Piers of Casks. In order to determine the number

of casks necessary to form a pier, the buoyancy of a cask must be calculated. This may be done by one of the following rules:—

(1) Find the contents of the cask in gallons and multiply this by $8\frac{1}{3}$; the result will be almost the *total* buoyancy in pounds.

(2) By the formula

$$5c^2 \, 1 - W = x$$

in which c is the circumference of the cask in feet half way between the bung and the extreme end; l is the length in feet, exclusive of projections, measured along a stave, and W is the weight of the cask in pounds; x being the *total* buoyancy.

If the cask is closed, $\frac{9}{10}$ of the *total* buoyancy equals the *available* buoyancy.

326.—To find the distance between two piers of casks: Find the available buoyancy of each cask. Multiply this by the number of casks in the pier. This gives the available buoyancy of the pier. To the weight per running foot that the bridge is to bear add the weight per running foot of the superstructure. Divide the available buoyancy of the pier by this sum; the quotient will be the required distance.

327.—In regard to *piers of casks*, the following should be noted:

(1) That piers of casks, when in bridge, should always be rigidly connected to each other at their ends by tie balks.

(2) That the tie balks should be lashed to both gunnels of each pier.

(3) That while the roadway balks may not be lashed to the gunnels and to each other, it should be done if there is much sway on the bridge.

328. –Piers of Open Casks. This is the simplest and most convenient method of using casks for piers, as it requires only a few nails and poles, dispensing with ropes which are sometimes hard to procure.

To make a raft of this kind, as shown in Pl. 44, Figs. 5 and 6, stand 10 or 12 barrels side by side, touching each other; nail 4 poles across the outside of the barrels, two at top, two at bottom, the nails being driven from the inside into the poles, which, as the heads are out, can easily be done. Place another row of barrels beside the row thus fastened together and nail them to the two poles of this row. Nail two poles to the outside of the second row of barrels, one at top and one at bottom; push the barrels thus connected into the water.

If too many rows are connected on land they will become too heavy to handle. Any number of rows, however, can be attached in the manner described above. When the raft is completed, the projecting ends of the poles outside are lashed together, and, at the points of contact of the barrels, a stout wire nail should be driven through and clinched.

329.—The total buoyancy of a cask may be calculated by the formula given above. If this should be 400 lbs., the safe load for smooth water would be at least 300 lbs.; that is, the available buoyancy is about $\frac{3}{4}$ the total buoyancy. A square raft of 10 such barrels to a side would carry safely 30,000 lbs.

330.—**Piers of closed casks.** The usual method of forming large casks into a pier is shown in Pl. 43, Figs. 1 and 2. The following are the successive steps in its construction:—

Stores required for a pier of 7 casks: 7 casks; 2 gunnels; 2 slings; 12 braces.

To build a pier of the foregoing stores, 1 N. C. O. and 16 men will be required. The detachment is marched to the site on which the material is placed and forms the casks into piers by the following commands and means, 4 men being detailed as gunnelmen and 12 as bracemen.

(1) **Align casks.** At this command, the casks are brought to the designated place by the bracemen and aligned, touching each other, bung uppermost.

(2) **Place gunnels.** At this command, the gunnels are placed on the outer ends of the casks by the gunnelmen.

(3) **Adjust slings.** At this command, gunnelmen bring up the slings and stand at the ends of the gunnels, the bracemen being opposite the intervals between the casks. The gunnelmen at one end place the eyes of the slings over the ends of the gunnels, and those at the other end secure the slings to the ends of the gunnels by a round turn and two half-hitches. The bracemen keep the slings under the ends of the casks with their feet. A sling is made of 1 in. rope and of sufficient length for an eye splice 1 ft. long, at one end.

(4) **Fasten braces.** At this command, the bracemen, having provided themselves with braces, pass the eye of the brace under the sling in the center of their interval, the end passed

through the eye and the brace hauled taut, the sling being steadied by either foot. The brace is then brought up outside the gunnel, directly over the eye and a turn round the gunnel taken to the left of the standing part.

(5) **Haul taut.** At this command, each braceman removes his foot from the sling and hauls up the standing part of his brace with his right hand, holding on to the turn with his left; as soon as the brace is taut, the turn is held with the left hand and the remainder of the brace in a coil is placed on the cask to the left.

(6) **Cross braces.** At this command, each braceman takes the brace of the man opposite him from the cask on his right, passing it between the standing part of his brace and the cask on his left, then back between his brace and the cask on his right, keeping the turn below the figure of eight knot on his own brace. The end is then placed on the cask on his right. Each man then takes back his own brace from the cask on his left, passes it under the gunnel to the left of the standing part, places one foot against the gunnel and hauls taut.

(7) **Rock and haul taut.** The bracemen, assisted by the gunnelmen, at this command, rock the pier backwards and forwards, the bracemen taking in the slack of their braces.

(8) **Steady.** At this command, the bracemen cease ~~hauling taut~~ rock and take a turn round the gunnel to the left of the previous turns.

(9) **Secure braces.** At this command, the braces are made fast by two half-hitches round the two parts of their own braces, close to the gunnels, drawing the two parts close together and placing the spare ends of the braces between the casks.

(10) **Turn the pier to the right and adjust sling.** At this command, the bracemen on the left side, assisted by the gunnelmen, turn the pier on its right side. The bracemen on the left side adjust the left sling.

(11) **Lower the pier, turn to the left, and adjust sling.** At this command, the bracemen on the left, assisted by the gunnelmen, lower the pier. The bracemen on the right, assisted by the gunnelmen, then turn the pier to the left. The bracemen on the right then adjust the right sling. The pier is complete.

331. Should the casks be very small, they may be put together as above described, forming small piers. These can then be united in one large pier by cross gunnels.

332.—Another method of forming casks into a pier is as follows:—(Figs. 3 and 4.)

Fasten the braces to a balk, two braces for each cask. Stretch out the braces perpendicular to the balk and lay the casks bung uppermost, end to end, on each side of the balk, each cask over its own braces. Upon the cask lay two gunnels, fastened together at the ends and one or two intermediate points by lashings, the distance between the gunnels being less than a bung diameter of a cask. Secure the braces to the gunnels by two round turns and two half-hitches. The lashings connecting the gunnels are then racked up. The two end gunnel lashings are lashed to the balk beneath the casks and these lashings are racked up taut. The pier is then complete.

333.—The barrels may be held in a *frame*, as shown in Pl. 44, Figs. 7 and 8.

334.—Piers of Logs. In order to determine the number of logs necessary to form a pier, the buoyancy of a log must be calculated.

To find the *total* buoyancy of a log. Multiply the solid contents of a log by the difference between the weight of a cubic foot of the log and a cubic foot of water.

335.—To find the solid contents of a log.

(1) Take a mean of the girths or circumference at the ends in feet and decimals. Square this mean and multiply it by the decimal .07956. Multiply this product by the length of the log in feet.

(2) Multiply the square of $\frac{1}{8}$ of the mean girth by twice the length of the trunk.

336. The weight per cubic foot of the timbers usually met with will be found in Chap. XV.

337. —Required the total buoyancy of a pine log whose mean girth is 6 ft. and whose length is 35 ft.

Applying rule 2, we have

$$\tfrac{6}{8} \times \tfrac{6}{8} \times 35 \times 2 = 100\tfrac{1}{8} \text{ cu. ft.}$$

$$100\tfrac{1}{8} \times (62\tfrac{1}{2} - 40) = 100\tfrac{1}{8} \times 22\tfrac{1}{2} = 100.8 \times 22.5 = 2,268 \text{ lbs.}$$

As lumber absorbs water, the *available* buoyancy is taken as $\tfrac{3}{6}$ the *total* buoyancy.

338.—To form a pier of logs. (Pl. 43, Fig. 5.) The largest and longest logs should be selected. Branches and knots should be trimmed off. The ends of the logs should be painted if the raft is to be used any length of time. The raft should be built in the water. Select a place where there is little current and where the bank slopes gently to the stream. Throw the timber into the water and moor it close to the shore. Note the natural position of each log in the water before putting it in the raft. The upstream end of each log should be drawn on shore and beveled to a whistle shape, so as to present less obstruction to the action of the current.

Arrange the timber in the position it is to have in the raft the butts alternately up and down stream, the up-stream ends forming a right angle, salient up-stream. The first log is brought alongside the shore and the end of a plank or a small trunk of a tree fastened with trenails or spikes to it about 3 ft. from each end. The log is then pushed off a little, a second log brought up under the transoms and in close contact with the first. The second log is then spiked like the first, and so on for each remaining log. Care must be taken to place the whistle ends up stream with the bevel underneath, and to spike the transoms perpendicular to the logs. If the stream is very gentle, the up-stream ends of the logs may be parallel to the transom.

Another method is to lash the logs together and fasten on the transoms with spikes or trenails. Or, lash the logs together and lash the transoms to the logs, tightening the lashings with rack sticks.

339.—Two additional transoms should be placed on the raft by whatever method employed in putting on the first. They should be the distance of the roadway or platform apart, at equal distances from the center of gravity of the raft, and bear upon all the logs. In order to obtain sufficient buoyancy, and allow sufficient water way, several courses of timber may have to be employed. For use in a bridge, a raft should have an available buoyancy of 15,000 lbs.

340.—If the raft is to be used as a *flying bridge*, it should have the shape of a lozenge. (Pl. 43. Fig. 8.)

341.—Anchors. Anchors for the U. S. Advance Guard Bridge Train weigh 75 lbs., and for the reserve train 150 lbs.

PLATE 45.

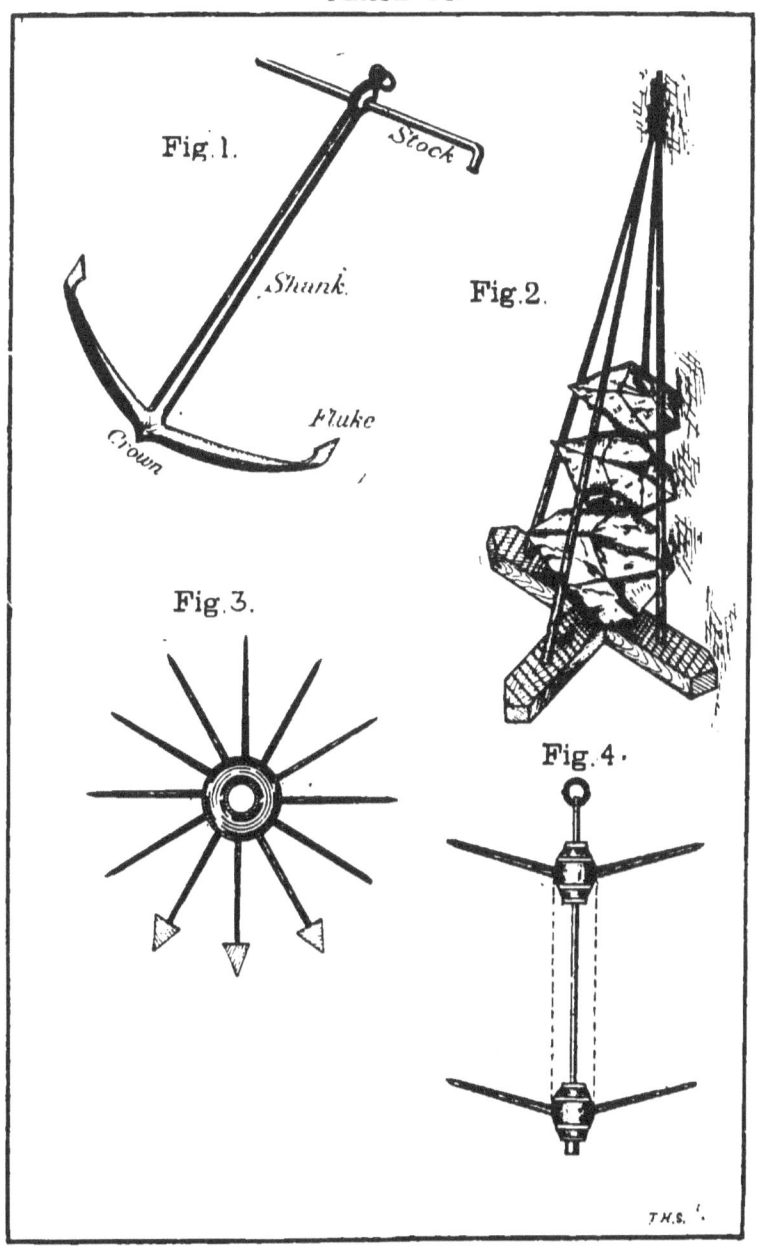

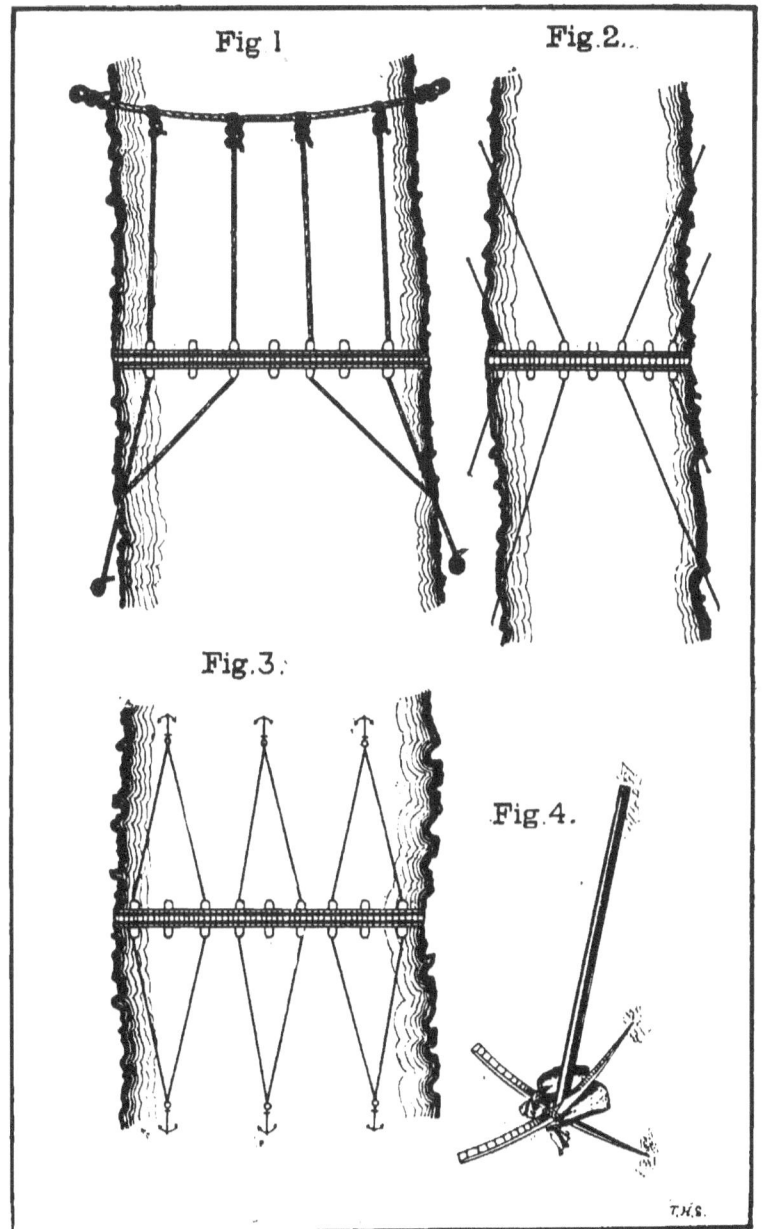

PLATE 47.

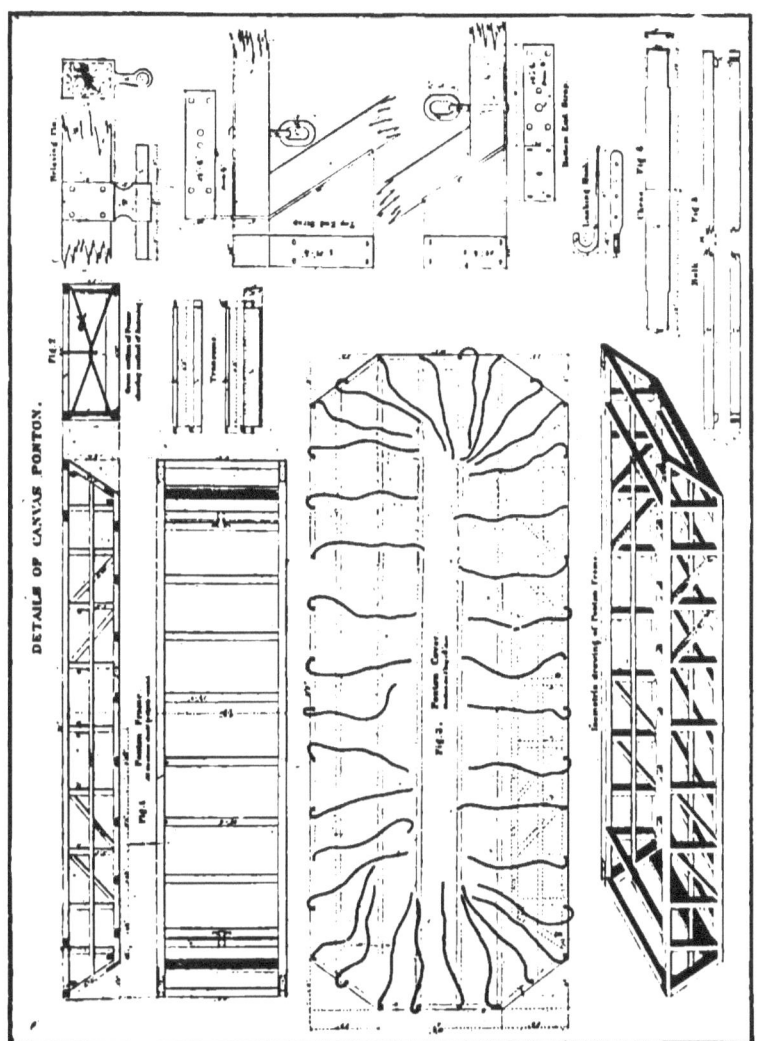

PLATE 48

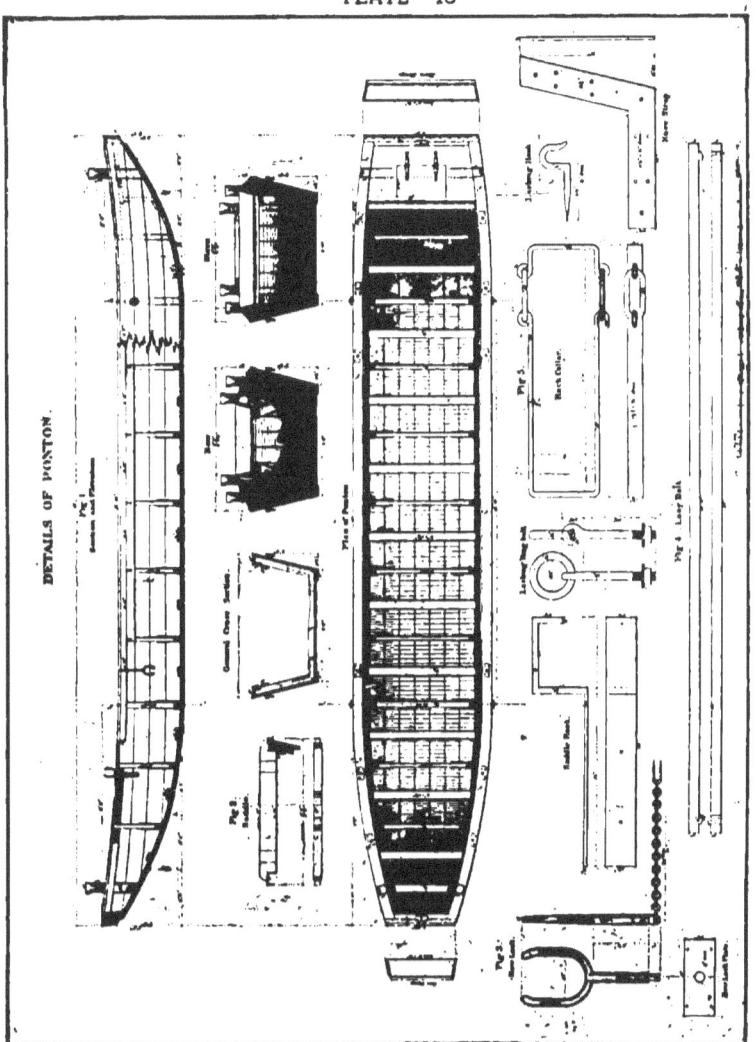

These will be sufficient for moderate streams. An anchor with the names of the various parts is shown in Pl. 45, Fig. 1.

342.—The distance of the anchor from the bridge should be at least 10 times the depth of the stream, otherwise the bow of the boat or ponton will sink too deep in the water. The direction of the cable must be the same as the current. The anchor cable should be of 1 in. rope and attached to the anchor ring by a fisherman's bend. A buoy might be attached to the anchor by means of a ½-in. breast line, in order to mark its position and serve as a means of raising it. The breast line is attached to the buoy ring by a fisherman's bend and round the shank of the anchor, close to the crown, by a clove hitch.

343.—The number of anchors will depend on the strength of the current. It is generally sufficient to cast an anchor up-stream for every alternate boat or ponton, and half that number downstream. If the stream is rapid, every boat should be anchored up-stream.

If *very rapid*, the bridge must be secured to a hawser, as shown in Pl. 46, Fig. 1. If the bridge is short, ropes can be stretched from the piers to the banks. (Fig. 2.) If anchors are scarce, one may be attached to two piers. (Fig. 3.)

Before being cast, the anchor should be well stocked. Rafts of casks or timbers bring a greater strain on anchors than boats or pontons.

344.—**Substitutes for Anchors.** One or two spare wheels with tires and felloes removed. (Pl. 45, Figs. 3 and 4.) Two or more pick-axes, laid together or fixed on one handle. (Pl. 46, Fig. 4.) A harrow with lengthened teeth, loaded with stones. Gabions filled with stones. Large stones or railway irons. Nets filled with stones. Frame filled with stones. (Pl. 45, Fig. 2.)

Care must be taken to allow the anchor to fall in good holding ground. For this purpose, a direction oblique to the current may sometimes be allowed.

345.—**Forming Floating Bridges.** Floating bridges may be formed in the following ways:—
(1) By successive pontons or boats.
(2) By parts.
(3) By rafts.
(4) By conversion.

346.—By Successive Pontons. (Pl. 49.) This may be done in two ways:—

(1) By adding to the head of the bridge, the tail being stationary. This method requires the roadway material to be carried an increasing distance. The men, however, do not have to work in the water.

(2) By adding to the tail of the bridge, the head, already constructed, being constantly pushed into the stream. The materials do not have to be carried so far as in the first case but it requires a number of men to work in the water and is not advantageous where the bank is steep.

In the first method, those boats or pontons which cast up stream anchors should be moored above the approach to the bridge, the others below.

347.—By Parts. (Pl. 49.) In this method, the boats or pontons are brought close to the shore above the bridge. For convenience in putting the parts together several chess are laid from the bank to the interior gunwale of one boat or ponton. The boats or pontons forming the part are then brought in place and balks placed on them. The chess forming the roadway are then placed on the balks, excepting a sufficient number at each end of the part to allow for the insertion of a bay between the parts. The parts, all constructed as directed, are then placed in position, each part carrying enough material to construct the connecting bay. The parts are joined with each other and with the abutment bay, which has been previously constructed.

348.—By Rafts. Each raft formed of 2 or more piers, is constructed complete and the rafts come into bridge in succession. Each of the methods, bridge by raft and bridge by parts, has the advantage of simultaneously employing a large number of men. (Pl. 49.)

349.—By Conversion. (Pl. 49.) In this method, the bridge is put together entire along the shore above the selected site. A tributary stream may be advantageous for this purpose. The bridge is then floated toward the site, care being taken to prevent the pivot end from touching the shore and the wheeling end from turning too fast.

350.—The various methods above described may be combined in the construction of one bridge.

PLATE 49.

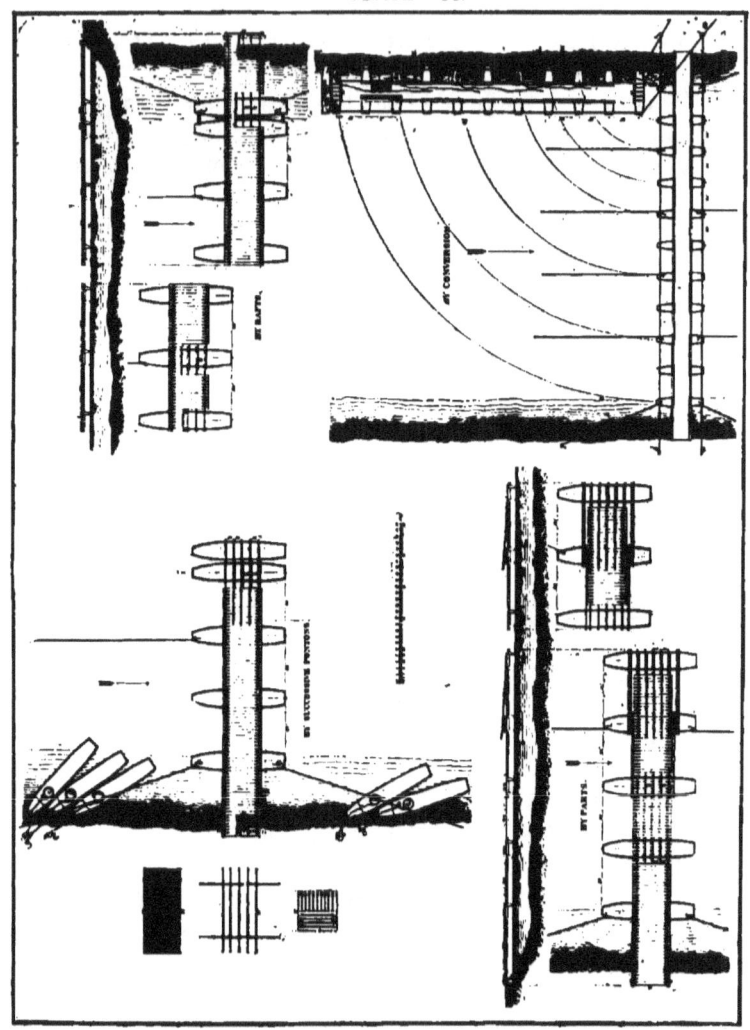

The connection of the bridge with the shore may be made by allowing the balks to rest on an abutment sill let about one foot into the ground, or by a trestle.

351.—If the stream is to remain open to traffic, it is well to have two or more rafts in mid-stream, arranged to swing so as to allow boats to pass, or the halves of the bridge may be swung for this purpose. Usually the passage is made by allowing the rafts or halves to swing with the current; they are then brought back against the current.

352.—**Floating objects.** Some arrangement should be made to protect the bridge from floating objects. This may be done :—

(1) By a guard of observation, stationed above the bridge, provided with boats containing anchors, grapnels, hammers, chains, etc. The object may be turned ashore, or, if this is not possible, an anchor may be attached to it to break its momentum.

(2) By a floating stockade, constructed of trees united by chains and forming a continuous barrier to floating objects. Its direction should be about 20° with the current.

(3) By constructing the bridge by rafts and withdrawing the menaced part, thus allowing the object to float past.

CHAPTER XVII. Roads.

353. —The frequent necessity, in the field, for the construction of a short piece of road, or the repairing of existing roads, makes it important that all who may at any time have this work in charge should be familiar with the principal requirements of it.

354.—Two desirable conditions in a road are that it be straight and level; where both cannot be obtained, straightness is sacrificed to levelness. Other things being equal, the length of a road may often be advantageously increased 20 ft. for every foot of vertical height avoided.

355.—Limiting Gradient. As levelness cannot always be obtained, various considerations fix limits for the steepness, called limiting gradients, which are to be used only when unavoidable; thus, for a very short distance, as an approach to a bridge, the limiting gradient may be $\frac{1}{10}$; a grade of $\frac{1}{12}$ should not exceed 100 ft.; one of $\frac{1}{15}$ should not exceed 200 ft.; $\frac{1}{20}$ should ordinarily be the limiting gradient for easy travel, while $\frac{1}{30}$ to $\frac{1}{35}$ is still better.

356.—Compared to what he can draw on a level, a horse can draw only about 90% on a grade of $\frac{1}{100}$, 80% on $\frac{1}{50}$, 50% on $\frac{1}{24}$, and 25% on $\frac{1}{10}$, but for a short distance he can exert 6 times his ordinary force.

357.—A road should, if possible, always rise continuously to its highest point and nowhere descend partially again.

358.—Width. For military purposes roads should be wide enough to allow wagons going in opposite directions to pass each other easily; this is usually taken at 16 ft. For wagons going in one direction only or with turnouts at intervals, and for infantry in column of fours, or cavalry in column of twos, 9 ft. will suffice, and for pack animals 6 ft. At turns in a zig-zag road up a hill the road should be level and the width increased from $\frac{1}{4}$ to $\frac{1}{2}$.

PLATE 50.

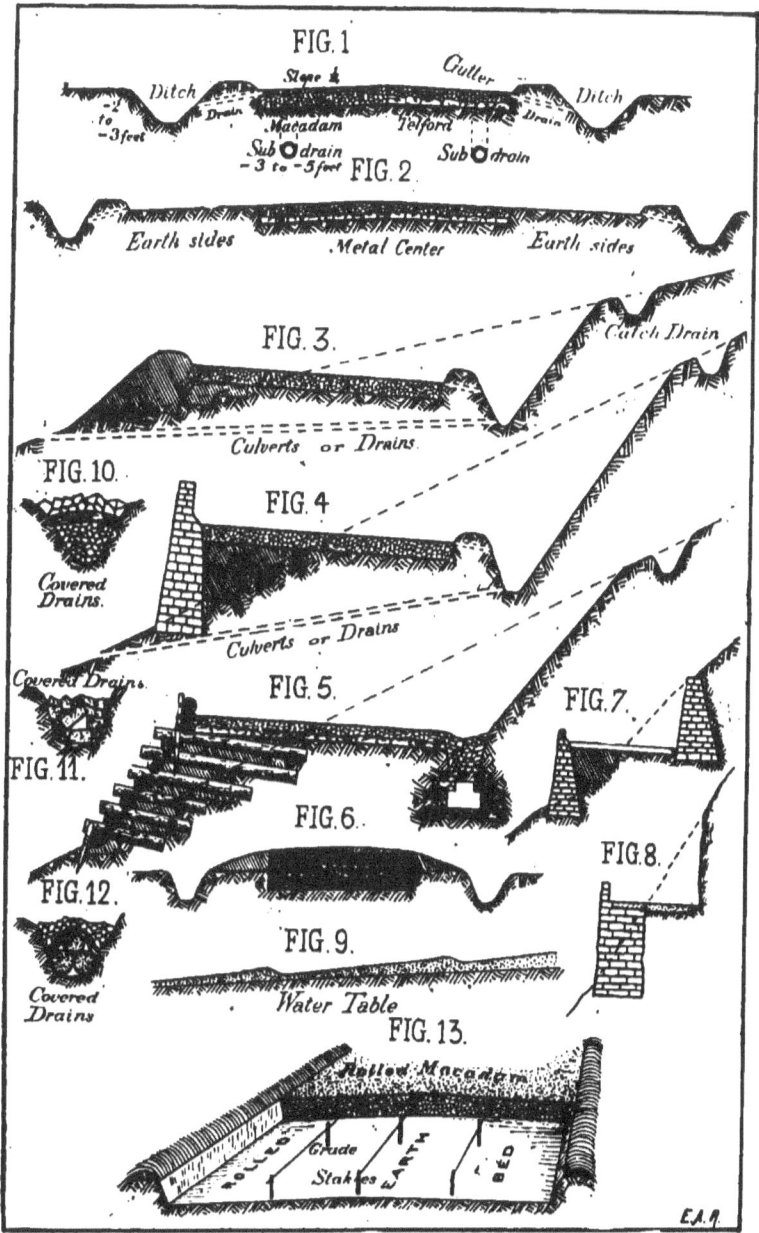

359.—Form. The best for the upper surface is that of two planes inclined at an angle of about $\frac{1}{14}$ and joined by a slight curve 5 ft. long. (Pl. 50, Figs. 1, 2 and 13.)

Between the road and ditches should be flat mounds raised 6 in. or more above the surface with sloping sides covered with sods or stone next to road, forming with roadway the gutters; they serve also to hold up the road material and as warnings at night of the proximity of the ditch.

On a hill side the surface should be a single plane inclined towards the hill. (Figs. 3, 4, 5, 7 and 8.)

360.—Road-bed. The surface of the road-bed should be dug out or built up and solidly compacted, either by rolling or ramming, and when ready to receive the road material should be of the same shape as the surface of the finished road, with shoulders at the sides to retain the material in place. (Fig. 13.)

On hillsides of gentle slope, the road-bed is usually made of half cutting and half filling, the lower side of the slope being stepped to retain the earth excavated, (Fig. 3); on steep slopes it is often necessary to both step the slope and build a retaining wall of stone, (Fig. 4); or of logs, (Fig. 5); or of other materials; on very steep slopes it may be necessary to build retaining walls on both sides, (Fig. 7); while in rocky formations the excavated hillside may be left nearly vertical, (Fig. 8.)

361.—Drainage. Nothing is of greater importance in road building than proper drainage. It is the life of a road. In a level country it is necessary to raise the road-bed to keep it always free from water. None must be allowed to remain on the surface and all must be drained from beneath. To accomplish this ditches must be dug on both sides of a road on level ground and in cuttings, from 2 to 3 ft. below the road-bed and of a width depending on the amount of water to be discharged. (Figs. 1 and 2.) In wet places, low-lying lands, clayey and springy soils, the ditches must be deeper and sub-drains 3 to 5 ft. below the road, emptying at intervals into the side ditches, must be made to keep it dry. (Fig. 1.)

Rain falling on the surface of the road is collected in the gutters on the sides and run into the side ditches by drains at frequent intervals.

On a hillside, between the road and the hill is the ditch, from which the water is discharged through culverts or covered drains under the road into the natural watercourses. Catch drains along the top of the cutting are made to prevent the slopes being washed down and the water from above finding its way to the road.

Where open ditches are liable to become filled, some kind of covered drain must be used. (Figs. 5, 10, 11 and 12.)

Theoretically, a road should be perfectly level, but for purposes of drainage, in the direction of its length, it should have at least a slope of $\frac{1}{125}$.

On a steep road, shallow paved water tables extending obliquely across the road are sometimes necessary to catch the water running down the road and carry it to the gutters, or small mounds crossing the road obliquely are substituted. (Fig. 9.)

362.—The surface of a road ought to be as smooth and as hard as possible, for which purpose various kinds of covering material are put on the bed.

As the roadbed must be kept thoroughly dry at all times by the ditches intercepting all ground water, so the stone or other covering must be so thoroughly rolled and compacted that no water falling upon the surface can possibly find its way down to the foundation and through it to the bed.

363.—When roads are made of broken stone the material in the Telford class is composed of two parts: the foundation and the covering. (Fig. 1, right half.) The foundation consists of a uniform thickness of not less than 5 in. of any durable broken stone with bases about 5 in. x 10 in. laid close together by hand, larger faces down, firmly wedged with smaller stones in the interstices, and the whole sledged and rolled to a uniform surface. Then a thin layer of binding material, as clay or loam, is sprinkled over it and rolled. On this is put the covering consisting of a layer of about 3 in. of broken stone of uniform, well shaped cubical pieces which will pass through a ring from 2 to 2¼ in. in diameter, and rolled to a uniform, compact surface. Then another layer of binding material is added and well rolled. Another layer of stones, 3 in. thick, of sizes from 1 to 2 in. in diameter is next spread and rolled as before. On this may be spread another binding coat, well rolled, then a thin layer of fine screenings or fine gravel free from dirt. Often, where traffic is light and expense large, a single layer of broken stone 4 in. thick is put on the foundation.

364. In the Macadam class (Fig. 1, left half) the hand-laid foundation is not used, but generally three layers, each from 3 to 4 in. thick, of broken stone and binding coats, as described above, are spread and rolled until smooth and compact.

For light traffic a single layer of 4 in. is sometimes used.

365. The best stone is a compact, fine-grained syenite, basalt or trap rock. Hornblend, actinolyte, dioryte, and some other rocks make good material. Quartz and flint, though very hard are brittle, difficult to work, and not so good. Granite, on account of mica in it, breaks up and grinds away too easily. Gneiss is poorer than granite. Slatey rocks generally break up too easily. Limestone, generally too soft, grinds away easily, making a very disagreeable dust. Softer stones may be used for the foundations and lower layers but only the hardest and toughest should be used for the coverings.

366.—Earth roads require even greater care in draining, grading, and forming the surface than those described, and a transverse slope, not less than $\frac{1}{10}$, to hasten the flow of surface water to the gutters. No sods or vegetable refuse should be allowed in grading or filling ruts, but gravelly earth, if obtainable.

Roads are frequently made with a metal portion in the center and earth roads, called *wings*, on the sides. (Fig. 2.)

367.—It is almost impossible to construct a road of clay which will be good in wet weather, but a very sandy road may be improved by working a little clay in it.

368.—For gravel roads the bed is first formed as described. The gravel is screened to remove stones larger than $2\frac{1}{4}$ in. in diameter and such as are less than $\frac{3}{4}$ in.; and all earthy matter. A layer of the screened gravel, 4 or 5 in. thick, is then spread and rolled, then another layer of 3 or 4 in. which should also be well rolled.

369.—Repairs. Ruts appearing should be immediately filled in, and traffic directed over all parts of road. Before spreading stones all mud should be cleaned off and the surface picked up a little to allow the new stone to bind into the old, wet weather being preferred, or the stones should be sprinkled. Ditches and culverts must be cleaned as needed.

370.—In crossing marshy ground that cannot be well drained, logs of suitable lengths laid side by side across the road, over which is spread a covering of earth or gravel, are sometimes used.

371.—Brushwood, made into fascines and hurdles, may be used the same way as a foundation. With fascines, the top row should extend across the road and be of a length equal to the width of road. (Fig. 6.)

372.—Where lumber is the cheapest material, plank roads may be built by first laying parallel rows of sleepers or sills flush with the ground, about 4 ft. apart, in the direction of the road, on which boards, 3 in. thick by 9 to 12 in. wide and 8 ft. long, are placed crosswise.

373.—The construction of communications to all parts of a position to facilitate the movement of troops, etc., from one part to another, is almost always a certain necessity. These would rarely be more than temporary, but, if made on the lines indicated, as far as time and requirements permitted, so much the better.

374.—Roads and paths may have to be cleared through woods; wet places made passable by corduroying or filling up with brush, fascines, etc.; and approaches made to ascend steep places.

Wherever roads cross or separate, signs should be put up telling exactly where each leads.

PLATE 51

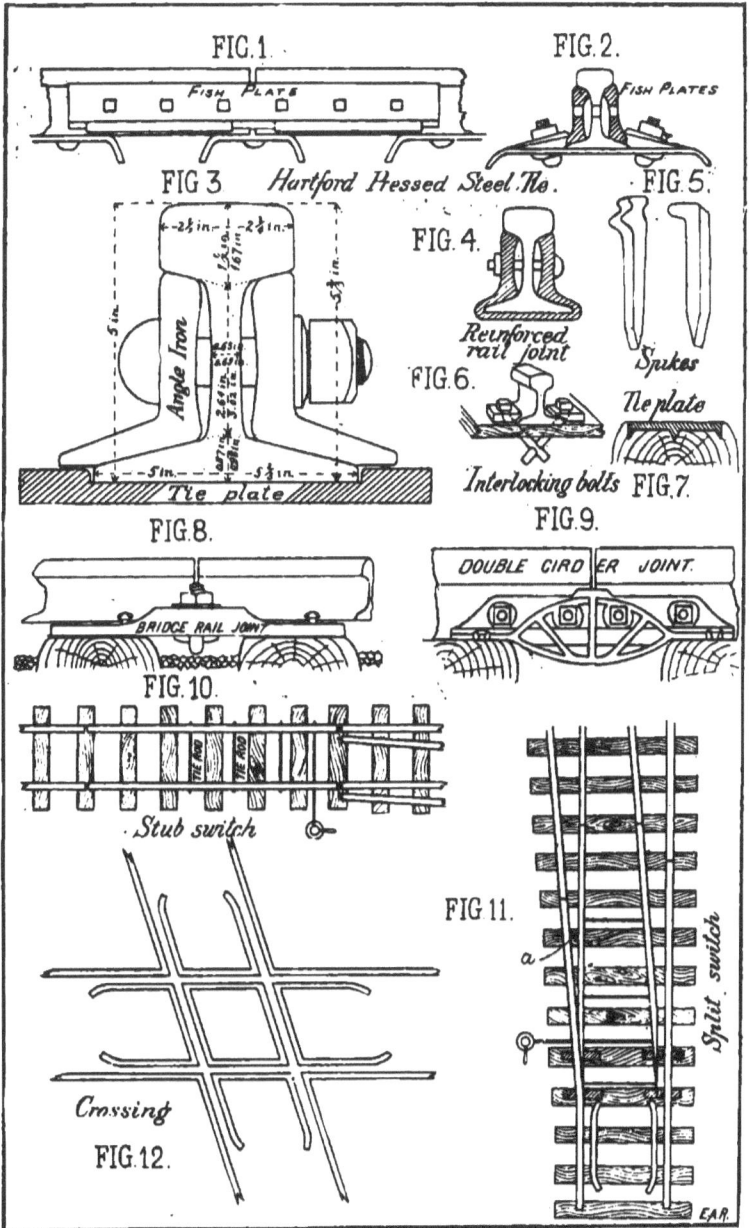

CHAPTER XVIII.—Railroads.

375.—In military operations, the principal duties of troops in connection with railroads will be either the repairing of lines that have been partially destroyed, or the destruction of lines to prevent their use by the enemy.

376.—A railroad, as existing in its completed form, will be briefly described to indicate the state to which it should be brought by repairs after destruction, and to so familiarize one with it as to suggest methods of most effectually destroying it.

377.—A railway line consists of a series of straight lines of different lengths, called tangents, which are joined by curves. The roadbed is first prepared with a smooth hard surface (sloping slightly from the middle to each side for drainage) from 10 to 12 ft. wide for a single track, and from 21 to 25 ft. for a double track. On this is placed the ballast, from 12 to 24 in. thick, of broken stone, gravel or cinders, etc., for the purpose of distributing the load over a larger surface, holding the ties in place, carrying off the rainwater, affording a means of keeping the ties up to grade line and giving elasticity to the roadbed.

378.—The ties are generally of wood, hewn flat on top and bottom, from 7.5 to 9 ft. long, 6 to 10 in. wide, and about 7 in. deep. It is customary to sink them about half their depth into the ballast. Their object is to hold the rails in place and furnish an elastic medium between the rails and ground. The distance apart is usually 2.5 ft. from center to center, but depends upon weight of engines and strength of rails. They should be uniformly spaced to distribute the weight equally.

Tie plates (Pl. 51, Figs. 3 and 7) are often used to prevent the rails from crushing into the ties.

379.—Tests of metal ties in the interests of economy and efficiency have been made with satisfactory results. On some level portions of the N. Y. Central R. R. are used the Hartford pressed steel tie (Figs. 1 and 2), to which the rails are fastened by clamps bolted to the tie.

380.—The form of rail used in the United States is shown in Fig. 3, being the "T" rail, which varies in weight from 12 to 100 lbs. per yard. The mean dimensions of 80 lb. rails are given on left hand side of figure and of 100 lb. rails on right hand side. They are placed 3 ft. apart for narrow gauge, 4 ft. 8.5 in. for standard gauge, while 6 ft. is the broadest gauge in the U. S., measured from inside to inside of head. The tops of rails must be slightly inclined to fit the cones of the wheels.

381.—The weak part of a track is at the joints. The old method of using chairs under the ends of rails has about ceased, the practice now being to fish the joints by plates (Fig. 1), and angle irons. (Fig. 3.) There are also used what are known as the Reinforced rail joints (Fig. 4), Bridge rail joints (Fig. 8), Double Girder rail joints. (Fig. 9.)

382.—Rails are fastened to the ties by spikes, the best being made with sharp, chisel-edge points, clean, sharp edges, and smooth surfaces, so as to cut and press aside the fibers of the wood, instead of tearing them. Attempts to increase the holding power by jagged or twisted spikes have been unsuccessful. On bridges, interlocking bolts (Fig. 6) are much used instead of spikes. To keep the track in the right line, allowance must be made for the contraction and expansion of the rails, by not placing them in contact at the joints, and the holes for the bolts must be elongated.

383.—The centrifugal force of a train passing around a curve tends to throw the wheels against the outer rails, which is partially counteracted by raising them to throw the center of gravity inward and cause the car to slide inward. Each rail in a curve ought to be bent to fit the curve before being laid.

384.—On single tracks, there are laid at occasional intervals short pieces of track, called sidings, to enable trains to pass one another. The arrangement for passing from one track to another is the switch, which consists of a single length of rails, movable at one end by a lever, so as to connect with either pair of rails. The simplest form is the stub switch (Fig. 10), which leaves one line always open while the other is continuous. The one in common use is the split or point switch (Fig. 11.) Various devices are used for locking and interlocking switches, to avoid accidents.

PLATE 52.

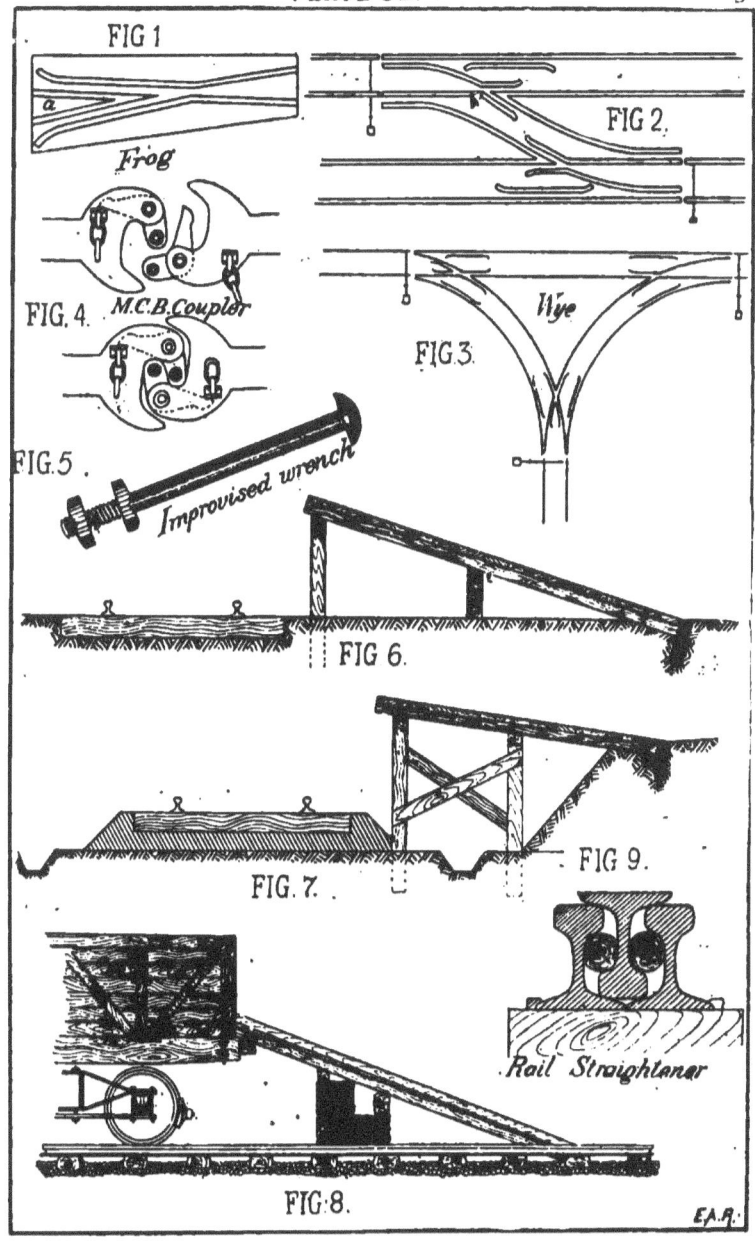

At the points where the inner rails cross is placed a frog (Pl. 52, Fig. 1), which enables the wheels to pass over the inner rail of the other track.

385.—Crossings occur where two tracks intersect, and consist of four frogs and corresponding guard rails. (Pl, 51, Fig. 12.)

386.—Where one main line passes to another is called a junction and the ordinary switch is used. In crossing from one track to a parallel track the rails are arranged as in Pl. 52, Fig. 2.

387.—A wye, from a similarity to the letter "Y," is an arrangement of tracks for turning around engines and cars and connecting cross-roads. (Fig. 3.)

388.—Turntables are platforms, turning on rollers upon an underground circular track, used to transfer engines and cars from one track to another and to turn them around.

389.—The locomotive engine is the power on railroads. They weigh from 58,000 lbs. to 190,000 lbs. without tender, and from 218,-000 lbs. for passenger to 310,000 lbs. for freight, with tender, all loaded, and draw 2,400 or more tons on a level. The amount of coal consumed being from 40 lbs. to 70 lbs. per mile run.

390. The rolling stock consists of passenger cars for about 60 persons, 48 to 52 ft. long, 9.5 ft. wide, weighing from 40,000 lbs. to 60,000 lbs.; sleeping cars for 64 passengers, 60 to 70 ft. long, 9.8 ft. wide, weighing 60,000 lbs. to 90,000 lbs.; mail, express and baggage cars, 45 ft. long, 9.3 ft. wide, weighing about 27,000 lbs.; freight cars consist of Box, Refrigerator, Hay, Furniture, Oil, Stock, etc., and are about 34 ft. long, 8.5 ft. wide, weighing about 20,000 lbs., capacity 20 tons; flat cars, 34 ft. long, weigh 16,000 lbs. to 19,000 lbs. Height of top of box cars above rails about 15 ft. Freight cars are being rapidly provided with the M. C. B. automatic couplers. (Fig. 4.)

391.—The buildings consist of passenger and freight depots, engine houses, fuel sheds, water tanks, repair shops, and section houses. At convenient points are generally located yards where stock can be loaded and unloaded. It may sometimes be necessary, however, to load and unload animals and supplies in the field along a railroad where there are no platforms or other conveniences, which must then be built.

392. A simple form of ramp, in the absence of anything better, could be made by taking 3 or 4 planks 3 in. thick, 10 to 12 in. wide, and 10 to 14 ft. long, fastening them together side by side, preferably by footholds nailed across on top, and several cleats on the bottom; otherwise, by lashing, wiring or by stakes at the bottom when in position, and wedges in the car door. The ends on the ground should be slightly sunken and rested against a cross beam. Ropes should be hung along the sides and blankets or canvas hung on them. Props of some kind, as sacks of grain, bales of hay, etc., can be placed under the middle to strengthen it if necessary.

393.—Another form of *portable ramp*, which could be carried on all railroad trains where they might be needed, consists of 6 long timbers 4 in. x 4 in. x 14 ft., 6 short timbers 4 in. x 4 in. x 6 ft., 24 boards 1.5 in. x 12 in. x 6 ft. with footholds nailed lengthwise on one side.

To load or unload horses, rest the ends of three of the long timbers, equally spaced, on the car floor, the other ends resting against a short timber, sunk in the ground and staked down. On these place the boards forming the floor; on each side of the ramp, on the boards, lay a long timber and fasten the ends to the timbers underneath. The boards should have cleats on under side to prevent slipping sideways. If necessary, some of the remaining boards can be set edgewise between posts of the short timbers as an intermediate support.

394.—*To unload a number of cars*, enough men can be placed under the ramp, near the car, to raise it high enough to allow the car to be removed and another run in place, thus avoiding taking the ramp apart for each car.

395.—*Semi-permanent platforms and ramps* may be made as in Figs. 6 and 7, if rails and boards are available.

396.—*To load or unload wagons and guns* from a flat car, place the ramp against one end (Fig. 8), using four long timbers for stringers on which the boards are placed, the other two long timbers being used for side rails. Support underneath with boards set on edge, held between some short timbers, or with bales of hay, sacks of grain or otherwise, as necessary. A couple of boards can be used to run the wheels from the car on to the ramp and others at the foot of ramp to carry the wheels across the rails. The lower ends of the stringers should abut against a tie, if possible; if not, they should be staked down.

397.—Pl. 53, is a design of a portable ramp devised by Major E. G. Fechet, 6th U. S. Cavalry. The ramp consists of 7 boards

PLATE 53.

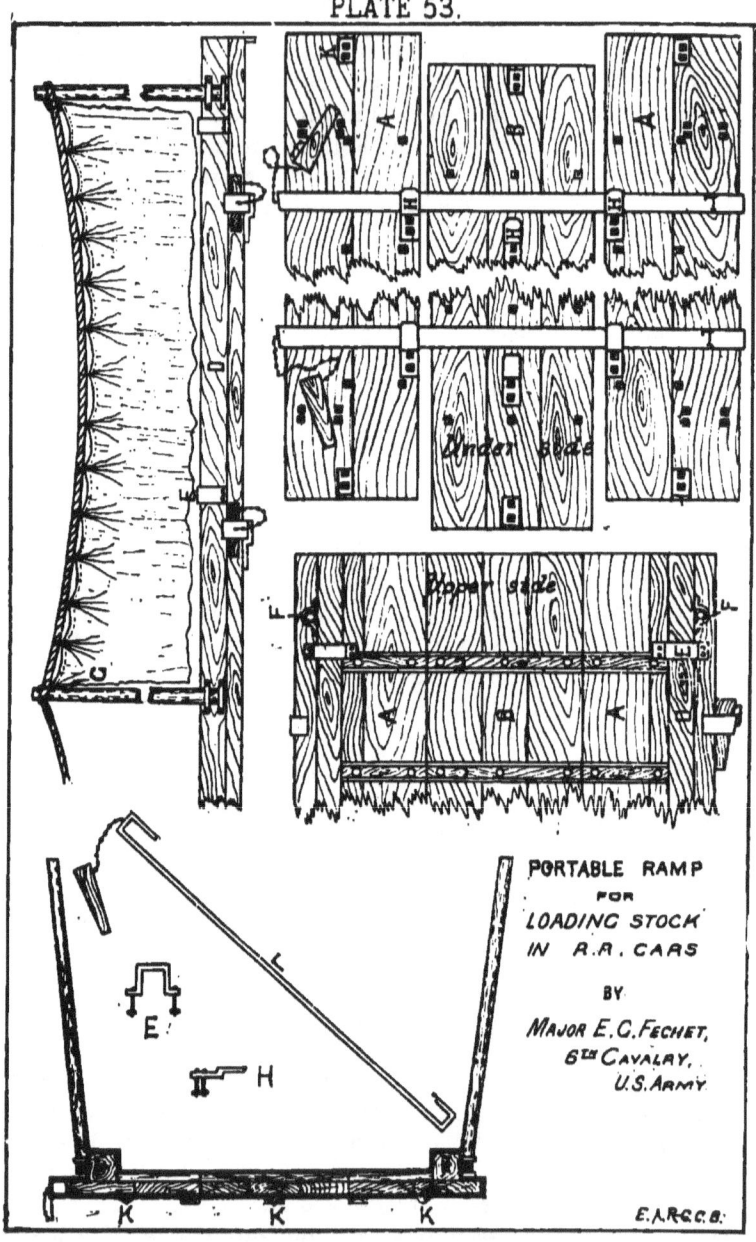

PORTABLE RAMP
FOR
LOADING STOCK
IN R.R. CARS

BY
MAJOR E.C. FECHET,
6TH CAVALRY,
U.S. ARMY

1.5 in. x 7 in. x 12 ft., joined together in three sections (2 for the outside, "A," "A;" and 3 for the middle one, "B"); by wooden strips "C," 1 in. thick, and 2 in. wide, bolted to the upper surfaces, 1 ft. between centers; these strips also serve as foot holds. Along the middle of the outside boards extends a side rail, "D," 3 in. x 3 in. held firmly by the iron straps, "E," ¼ in. x 2 in. On the outside of each side rail are 3 sockets "F," for standards "G" 3 ft. high, along the tops of which are to be stretched ropes or chains from which canvas or blankets are hung. On the under side of each section 3 ft. apart are bolted iron cleats, "H," 0.5 in. x 2 in., beginning at 18 in. from the ends. On the ends of each section are bolted iron claws, "K," for catching the car floor or door slide, to prevent slipping when in position for use. The three sections are held together for use by 4 iron tie bars, "L," 0.5 in. x 2 in. which are placed under the cleats "H," and the whole firmly keyed as shown. This form of ramp may be made longer or shorter, narrower or broader, as desired. By taking out the standards it may be hung on the side of a car between a door and end. It is easily taken apart and transported in a wagon, and as easily put together when needed. It is designed to combine both strength and lightness. It weighs about 400 lbs. complete.

398.—Disabling and destroying railroads. Under the head of *disabling* will be mentioned means, the effects of which will only temporarily interrupt traffic, can be easily repaired, but will cause delays.

399.—Under *destroying*, such as are more serious in their effects: either causing extensive repairs or a change of route to avoid them.

400.—The disabling of railroads will usually be done by raiding parties of cavalry, while the destroying of them may be done by such parties or specially detailed troops trained for such service.

401.—It must be understood that no railroad is to be destroyed except upon the orders of the officer commanding in the field. If otherwise, and it should be taken from the enemy, the damage done might seriously embarrass future operations. Before ordering any destruction the quesiton will arise—" Is destruction absolutely necessary?" "Will it be of no further use and is every hope of regaining it gone?" "Are the advantages to be gained sufficient to compensate for the damage that will be done?" All the attending circumstances should be carefully considered, especially if in one's own country. The choice of points for destruction and the most effective means are subjects for study. It is useless to destroy anything that will not seriously embarrass traffic.

402.—*A railroad may be disabled* by removing rails at various intervals, then destroying or hiding them; or, if a large number of men are at hand, select a high embankment, line the men along on one side of the track, disconnect the rails at each end of the line of men, then, at a signal, they raise the track on edge and let rails and ties together go over the embankment. Thus treated, rails and ties must be separated before being replaced. An improvised wrench for removing nuts on fish plates is a bolt with two nuts on it, just far enough apart to grasp the nut to be removed. (Pl. 52, Fig. 5.) If time is an object, remove outside rails on a curve, or disconnect a joint on each side and throw them as a switch to derail the train either on an embankment or in a cut, or use explosives as described in Chap. XX.

By laying rails across a pile of burning ties until red hot in the middle they may be easily bent around a tree or telegraph pole; they may be twisted by heating, as above, then using bars or pick axes placed in the holes in each end and working in opposite directions.

They may be torn from the ties and twisted cold by using Gen. Haupt's "U"-shaped rail twister, shown on Pl. 40, Fig. 8. Ten men with two twisters, two axes, two stout pieces of rope 35 ft. long, can tear up and twist a rail in 5 minutes. The junctions of lines are important points to attack to disable a track.

Water tanks may be rendered useless for a time by breaking holes in them, removing pistons from pumps, etc. Fuel, ties, and small bridges may be burned. Engines may be disabled by burning out the flues, removing or breaking different parts of the machinery, filling suction pipes of pumps with waste, etc. Cars may be disabled by removing couplers, axle boxes, breaking or removing trucks, etc. The use of mines under the track, so arranged as to be exploded by the passing of trains, is an effective method of interrupting traffic and shaking the morale of troops being transported.

403.—*To destroy a railroad*, if time is sufficient, remove rolling stock, rails, etc., to the rear. Otherwise, destroy large bridges, if of wood, by burning, using oil if it can be obtained, or by explosives, as in Chapter XX. If of iron, steel, or masonry, by explosives, as in Chapter XX. If there are tunnels on the line, select longest ones and blow them in at as many points as possible, or cause two wild trains to collide in the middle, afterwards blowing in the ends. Those with sandy soil are the best. Deep cuttings with retaining walls may be filled in by use of explosives. If trees, poles, wires, etc., can be mixed in—so much the better. Blow up tanks and engines, burn all fuel, cars, repair supplies, etc. Fire a cannon ball through engines.

404.—The repair of railroads will best be accomplished by a construction corps having some of the elements of permanency in its organization; or, at least, by small squads of experienced men to which others could be added by temporary detail, whenever active operations require such increase. They should be established as near to where their services may be needed as possible.

405.—Bridges should, in the beginning, be classified and numbered, so that a single reference to the class and number will give complete information as to its character, dimensions, etc. At designated points will be kept on hand, already prepared for putting in bridge, suitable materials for the repair of each class. This was done by the Union Army from 1861 to 1865, so that, when word was received that a certain bridge had been destroyed, by a reference to the class and number the reconstruction corps started out carrying with it just what was needed to repair the bridge. Even complete trusses for the larger class of bridges were prepared and kept ready for use.

406.—Tunnels and cuts which have been filled up can generally be cleared only from the two ends.

407.—Rails, fish plates, spikes, ties, etc., will be kept in store at secure places, for repairing any portions of destroyed track. Rails which have been simply bent can be straightened by various means. Gen. Haupt's method was as follows: Two ties were placed on the ground, across these two more ties and on top a single tie which was cut across one-half the depth of the rail to receive it and prevent it turning. Weight was applied at the two ends of the rail by men bearing down on poles placed thereon. The rail being moved back and forth until straightened, requiring from 4 to 5 minutes. Rails which had been heated and bent to a very sharp angle required more time, necessitating reheating and hammering until straightened. For this purpose, at special points were prepared furnaces consisting of two parallel walls of brick, stone or clay, with a kind of grate. The straightening table consisted of a large, square timber as long as a rail, on which were securely fastened three rails, as in Pl. 52, Fig. 9 on which the heated rail was laid and hammered until straightened. Twisted rails require re-rolling before they can be again used.

PLATE 54.

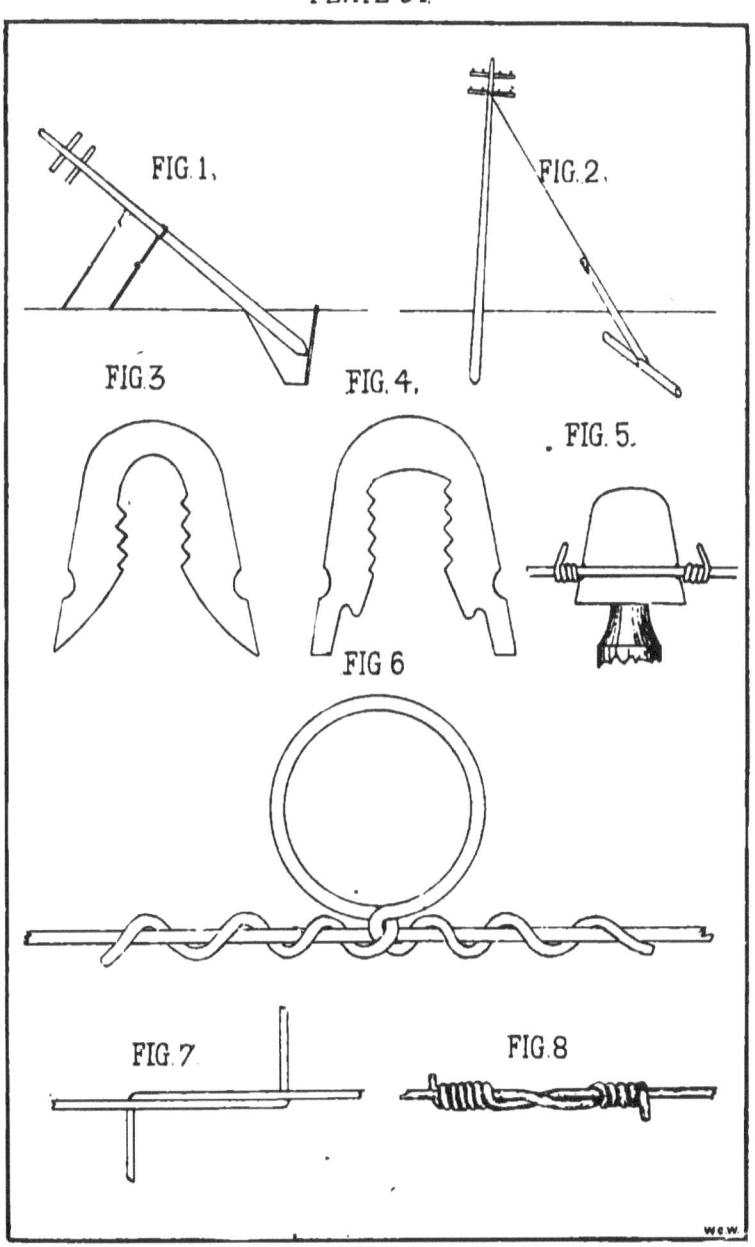

CHAPTER XIX.—Telegraph and Telephone Lines.

408.—In order that telegraphic messages may be sent from one point to another, it is necessary that there be a continuous metallic conductor from the first to the second point and that this conductor be insulated from contact with the ground or with anything leading to the ground. The conductor used in constructing permanent lines is of galvanized iron wire, generally of size No. 9. In military lines it is generally somewhat smaller on account of the weight. The wire is carried on poles and tied to glass insulators which are attached to the poles.

409.—Poles should be not less than 22 feet in length nor less than 7 and 5 inches in diameter at the larger and smaller ends respectively and should be stripped of bark and pointed at the upper end. The holes for poles should be not less than $\frac{1}{5}$ the length of the pole in depth. The poles should be raised as shown in Pl. 54, Fig. 1 and held vertically while the excavated earth is thoroughly tamped in from bottom to top; after the hole is completely filled the earth should be made into a small mound so as to shed water.

410.—When the brackets are attached to the pole directly, a seat should be cut in the pole with a hatchet and the bracket should be nailed on, using 1 twenty-penny and 1 forty-penny nail. Where the poles are intended to carry several wires, cross-arms are bolted to the poles, fitting into seats cut for them. The arms carry brackets not less than 15 in. apart. The arms should not be less than 20 in. from one another. Not less than 25 poles to the mile should be used, and in special cases the number may be increased to 30.

411.—Every 5th pole should be protected from lightning discharges by having a piece of line wire run from about 6 in. above the top of the pole to the ground. This wire must be so arranged that it cannot come in contact with the line wire should that become unfastened. Poles should be vertical except when necessary to incline them to resist strains, when they will be set at a slight inclination, in such manner that the component of the strain in the direction of the length of pole will tend to press it into the ground. Where exposed to great strains, or to continuously high winds, it may be necessary to guy the poles: this is done with stays consisting of two or more line wires twisted together and fastened near the top of the pole, the ground end being attached to a section of a pole or timber suitably anchored in the ground, as shown in Fig. 2. Where possible, the line of poles should be run on one side of the road and far enough from it to be safe from accidental damage by passing wagons. Where roads have to be crossed, the wire should be carried over on high poles so as to clear any possible wagon load.

412.—The insulators in common use in this country are of glass and of the form shown in Fig. 3. The one shown in Fig. 4 is preferable, as it is not so liable to cause leaks on account of moisture accumulating and forming a connecting film to the bracket and from that to the pole.

413.—The wire is attached to the insulators by pieces of wire called ties. These are generally of the same wire as the line. They are annealed and formed on an insulator and cut long enough to embrace the insulator and project 3 or 4 inches beyond the line wire.

414.—To hang the wire. The wire is carried up to the top of the pole and the lineman places a tie on the insulator, the line wire against the insulator above the tie wire, and bends the ends of the tie wire upward so as to sustain the line wire. The line wire is then strained by the lineman, either by means of hand power or by use of the wagon carrying the reel. When the ine wire is stretched so that it sags but about $1\frac{1}{2}$ ft. in 70 yds. the tie wire is wrapped around it about one and a half times, finishing with the ends of the tie wire pointing towards the insulator; this secures the line and completes the work. (Fig. 5.)

415.—In open country the line wire is strung on the insulator on the side towards the pole, so that, if it becomes accidentally undone, the wire will not drop. If in timbered country, then hang it on the side from the pole, so that when trees, etc., fall against the wire it will simply tear it away from the insulator, but will not break the line wire. When necessary to hang the wire on trees, a regular tree insulator should be used, and in default of this, the tie shown in Fig. 6 may be used, the ends being wound loosely so as to allow of an easy lateral motion to accommodate the swing of the tree. The poles should be numbered at each mile so as to aid linemen to report location of breaks and repairs.

Streams are crossed by hanging the wires on strong, high supports, taking care not to strain the wire so much as to cause it to break.

416.—The description of instruments and batteries, their connections and care, will be found in the Manual published by the Signal Service of the Army.

417.—Joints. Where wires have to be joined to preserve the continuity of the metallic circuit, the best joint is the American twist joint. To make this, clean the wires for a length of 5 or 6 inches, make a right angle-bend in each wire about 4 inches from the end, now join the wires so that the ends project on different sides and clamp both wires with a hand vise, then with a splicing iron turn the ends around the line wire, making the turns as close as possible; after the entire end is turned around the line wire, cut off the projecting end and dip the joint into melted solder; this protects the joint against rusting. The details of this joint-making are shown in Figs. 7 and 8.

418.—Military lines are generally of the kind designated as flying lines, *i. e.*, they are intended to accompany the army in the field, are constructed quickly for temporary use, and are as quickly dismantled and taken up. The poles used are small poles called lances, each about 2⅜ in. in diameter and 17 ft. in length, placed 2 ft. in the ground, and run about 40 to the mile. The batteries, line lances, and instruments are carried in wagons which accompany the army. A detailed description of the telegraph with directions how to erect and dismantle is found in the Manual of Signals for the U. S. Army. The ordinary telephone receiver (with magneto call bell) is used on the military lines; but for the use of outposts, reconnoiterers, and scouts, a special form of telephone cart and wire has been adopted, the following description of which is taken from the Report of the Chief Signal Officer of the Army, 1892:

"The frame of this cart is constructed of bicycle tubing, and 30 in. bicycle wheels with heavy cushion rubber tires are used. The cart is fitted with an automatic spooling device for reeling up the outpost cable. This device was made by F. S. Cahill & Co., and is a success. The cart carries 5 reels of cable and 1 reel knapsack for use in places where the cart cannot penetrate owing to underbrush, etc. As the extreme width of the cart, measured at the wheels, is only 26 in., it can follow any ordinary path through underbrush. The weight of the cart complete with spooling device, but without the reels, is only 53 pounds; when loaded with reels and reel knapsack the total weight is 157 pounds. The cart is well balanced upon its axle by a device which permits the point of support to be changed to balance the cart, as the distribution of the weight is changed by the cable being run out. In connection with the reel cart a telephone kit is used, and by attaching the double connector of the kit to one on the frame of the cart the telephone is kept in circuit and conversation can be kept up with the home station. The cart with its load can be easily drawn by one man, and by its use it will be possible to connect outposts with the main guard, or brigade with regimental headquarters, or brigade with division headquarters, in a few minutes of time. The experience of the English in Egypt has proved the value of the field cable line in action, as by means of these lines the Commanding General was kept in communication with different divisions of troops and with those actually engaged in the firing line. It is proposed to fit shafts to the cart so that a horse can be harnessed to it, thus securing great rapidity in running out the cable. The cart carries $1\frac{3}{4}$ miles of cable which can be paid out as fast as a man moves with the cart, and by means of the reeling apparatus and spooling device can be recovered at the rate of 4 miles per hour, or as rapidly as a man can walk with the cart."

419.—**Faults** are generally of three kinds—breaks or disconnections, leaks or escapes, and crosses or contacts.

Breaks occur when the metallic circle is broken or cut so that either the disconnection is complete, as when entirely severed; or incomplete, when partially cut or where a joint is rusted so much as to increase the conductive resistance. In these cases the instruments will work weakly or fail entirely.

Leaks, or escapes. Where the insulation is destroyed or is defective or where a wire comes in contact with a conductor to the earth or with the earth itself, a portion of the current leaks or escapes. When the wire is swinging the leak will be intermittent; when constant leakage is going on the instruments will work weakly; failing altogether when the leak becomes complete, then it is called a ground.

A cross occurs when two wires, each carrying currents, are brought into contact; thus the instruments on one line will interfere with the workings of those on the other. Generally occurs from parallel wires being swung over one another by the wind, or having a good conductor fall so as to touch both wires.

420.—Telegraph lines should never be damaged or destroyed, except in obedience to direct orders. Faults may be made by connecting the wires together with small wire (this makes a bad cross), or they may be connected with the lightning rods on the poles, thus running them to the ground.

When an office is taken, the instruments should all be disconnected and destroyed or taken away; the ends of the wires should be tied together. The batteries, if any, should be disconnected. To destroy the line, cut down the poles and burn them and cut the wire into small lengths. Subaqueous lines should be brought up with a grapnel and a piece cut out and cut into small pieces and thrown back into the water. Subterranean lines are generally laid in conduits, and at regular intervals man-holes are built to allow of repairs; the line may be detected by these man-holes, the conduit destroyed, and the cables disconnected.

CHAPTER XX.—Demolitions.

421.—In military practice, demolitions must be made with the least possible expenditure of time and explosive. At the same time, a charge which in itself seems large for a particular case may prove economical, in that it errs on the right side and a repetition of the work is not made necessary.

Military engineers have to destroy bridges, houses, walls, etc., render roads impassable, fell trees, and place mines. In all these operations, high explosives are generally used, on account of their portability, ease of handling, and great destructive effect. In some cases, ordinary gunpowder is available, and may be used to advantage if time allows of its proper placing and tamping. Guncotton, however, is the standard explosive for military work, and all formulas will be calculated for its use.

422.—Guncotton, as made at the U. S. Naval Torpedo Station, is in blocks about 3 in. square and 2 in. thick, each block being perforated to allow the insertion of the primer, and when dry the blocks weigh 10 oz. When necessary to use blocks of smaller dimensions, the large ones may be cut, when wet, by using a saw or sharp knife, care being taken to place the block to be cut between two boards, so that it will not flake or crack during the operation. Wet guncotton can be detonated by means of a primer of dry guncotton. Guncotton will absorb about 30% of its weight of water, and, when in this condition, it is comparatively safe, as it can only be ignited by fire and is difficult of detonation. When packed for transportation, the blocks are placed, while wet, in a tin can, and sealed hermetically; or, the can is left so that the water can be replaced when it has evaporated.

423.—When both are well tamped, guncotton has an explosive force from 2 to $2\frac{1}{2}$ times as great as gunpowder, but when no tamping is used it has a force 4 times as great. When guncotton

PLATE 55.

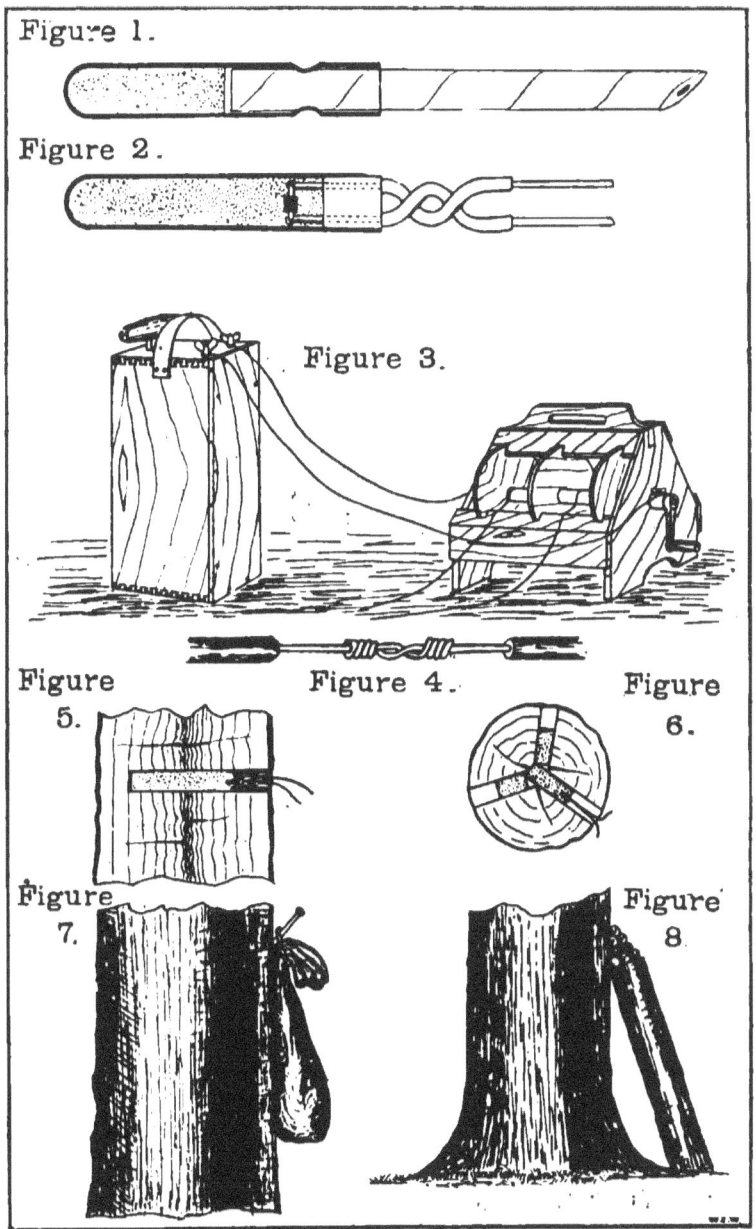

is needed for use as primers, a sufficient quantity is taken from the wet case, placed in the sun, and allowed to dry. In operations in the field, a small quantity of dry guncotton should be carried, so as not to waste time in drying the primers. Wet guncotton detonates with much greater force than dry, but it is necessary that this action be set up by the detonation of a dry block placed in intimate contact with the wet charge. The manner of firing guncotton is as follows:—

After placing the explosive in as close contact as possible with the object to be destroyed, and seeing that the blocks composing the charge are in intimate contact, the detonator is placed in the hole in the center of the dry primer block and the fuse is lit or the current sent through the wire, as the case may be. The detonation of the dry primer will set up detonation in all the blocks and the detonation will be simultaneous throughout the entire mass. When used in holes, guncotton should be dry, as wet guncotton is difficult to detonate under such circumstances.

424.—Detonators. The detonators are copper tubes, about ¼ in. in diameter, closed at one end, and partially filled with a fulminate composition which is ignited either by a fuse or by electricity. The ordinary blasting cap (Pl. 55, Fig. 1), or detonator, is intended for use with a fuse, and is designated as single, double, or triple force, according to the amount of fulminate composition used. In all military work, it is better to use the triple or higher force caps, as their action is sure, even on comparatively low explosives.*

425.—The fuse generally used in this country to ignite these detonators or caps is that made by Ensign, Bickford & Co., Simsbury, Conn., the grade known as "double taped" being the best for general work when the fuse is not exposed to prolonged immersion in water or damp ground. When it is necessary to use fuse for submarine explosives, the waterproof fuse should be used. The rate of burning of the fuse should be found by experiment before using. This is done by taking several pieces, 1 ft. long, and finding the average rate of burning and taking this as the standard. The rate at which this fuse is *intended* to burn is 3 ft. per minute, but it varies somewhat, so that when great nicety is required it should be tested as above.

---

* Single force caps contain 3 grs., double force caps 6 grs., and triple force caps 9 grs. of fulminate of mercury.

426.—To prepare a fuse and detonator for use, cut the fuse to the length required, leaving a square end; insert this end in the detonator until it rests against the fulminate, taking care not to scratch the fulminate. Then crimp the copper against the fuse so as to hold it firmly : this is done by means of pincers made for the purpose; or, in case these are not available, any pincers or the edge of a dull knife may be used, being careful not to crimp on the portion of the cap containing the fulminate. When ready to fire, insert the detonator in the hole in the center of the priming block, secure it by tying with wire or twine, and, when ready, light the fuse.

To ignite powder, the fuse alone is used, being placed so that the end is well centered in the mass of the powder and so secured that it cannot pull out.

427.—Electrical fuses and detonators are constructed so that ignition takes place upon the passage of a current of sufficient intensity through a platinum wire, around which is wrapped some fleecy guncotton or other inflammable material which is ignited by the platinum wire being heated to redness when the current is turned through the primer, this ignites the rest of the composition which causes detonation or explosion as the case may be.

Two forms are shown—the commercial (Pl. 55, Fig. 2) and the service (Pl. 57, Fig. 10.) The commercial is so constructed that it can be used with maximum effect only in detonating compounds; while the service fuse, by removal of the copper cap containing the fulminate, is converted into a simple electric fuse and may be used to ignite gunpowder mines. The current of electricity is generated by means of a battery consisting of several cells or by electrical machines of various forms. The means now generally used, and the one that gives surest results, is the magneto machine. On account of its portability and compactness, and because of its simplicity, the "Laflin & Rand Exploder, No. 3," is recommended for all work when electricity is to be used as the igniting agent. (Fig. 3.)

428.—When more than 5 fuses are to be fired in one series, it is advisable to use a larger machine of the same construction. In favorable cases, the machine in question will ignite as high as 12 charges, but, in military work, it is best to leave nothing to chance, and the limit stated above is a good one. The machine is cased in wood and its dimensions are 13 x 8 x $5\frac{1}{4}$ in., and

PLATE 56.

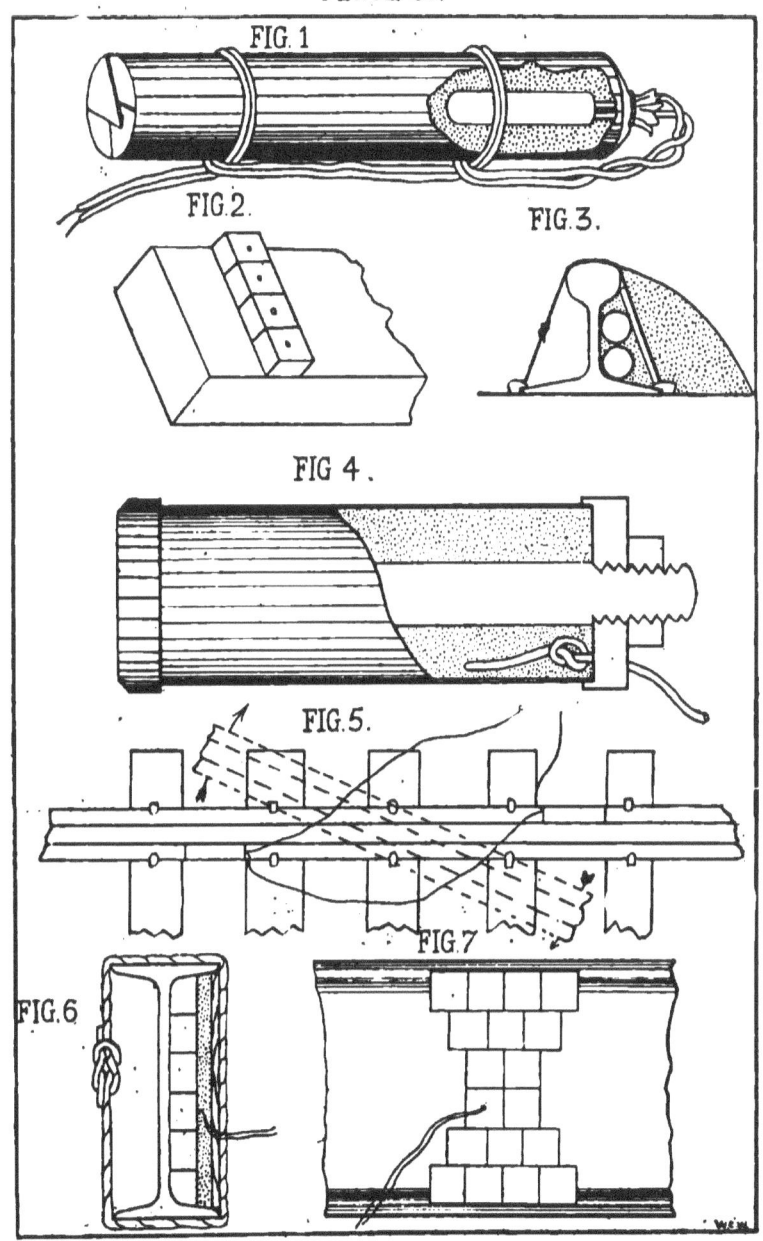

weighs 18 lbs. The connecting wires are attached to two brass binding posts, which are found on the top of the box. The handle on top of the box is lifted, withdrawing the ratchet bar to its full length, and, when the time arrives to fire the charge, the bar is pushed vertically downwards; moving slowly for the first inch or two, then by a swift, but even pressure till the lower end is stopped at the bottom of the box.

In all connection of wires, at least 2 in. of each wire must be cleaned bright and well wrapped around one another, as shown in Pl. 55, Fig. 4; under no circumstances simply hook wires together.

429.—After wires are joined, the joint should be insulated by winding rubber tape, or wide rubber bands, around so as to overlap the next previous turn. When water has to be encountered, wrap with rag or a strip of linen and cover with tar; or, use rubber tube as shown: the tube in this case is placed on one of the wires, and, when the joint is made, is pulled over it and tied tightly to the wires on either side of the joint. (Pl. 57, Fig. 9.)

Where dynamite is used, a hole is made for the detonator in the end of the cartridge with a sharp stick or lead pencil and the detonator with fuse or wire attached is inserted; the envelope is then tied around the fuse, so that the detonator cannot become detached from the cartridge, care being taken to place the detonator in cartridge only about ¾ its length, so that the charge may not be ignited by sparks from the fuse before the detonator is exploded, and all is ready. When electricity is used, push the detonator in its full length and take two half hitches, with wires, about the cartridge, so as to hold everything in place. (Pl. 56, Fig. 1.)

In all cases, where more than one package or piece of guncotton or dynamite is used, the closest contact possible should be made between the explosive and the material to be destroyed, and also between the different pieces of explosive. In all cases wherever the magneto machine is to be used, be careful not to connect the wires with machine till every one is at a safe distance from the place of explosion. When more than one charge is fired, the wires leading to the several charges should be connected as shown in Pl. 57, Fig. 6.

430.—**To fell trees,** bore a hole at the height desired, insert the charge, and fire, care being taken that the center of the charge is about the center of the tree. If the charge be of a length equal to or less than the diameter of the tree, then the hole may be bored directly through; but if greater, then two or three holes, intersecting at the center, must be bored.—Charge in ounces may be calculated by the formula $c = 6d^2$, in which c is charge in ounces and d the diameter in feet. To make the tree fall in a given direction, tie a rope to that side and pull. When time is not available to bore a hole, make a necklace and bind around tree.—Charge in ounces $= 48d^2$. When impracticable to make necklace, place charge in a bag, or support by stakes, so that all the explosive is against one side of tree, and fire.—Charge in ounces $= 60d^2$. In this case, the tree will generally fall towards the side of explosive. This method of felling trees is more expensive than with the axe, and is not resorted to except when time is not available for the slower method. It is not advisable to use either of the last two methods when it can be avoided, as there is great waste of explosive. (Pl. 55, Figs. 5, 6, 7 and 8.)

431.—**To destroy bridge timbers**, if rectangular, the charge is placed across the whole width of the beam; 10 ozs. will destroy a beam 12" high and 6" thick. The charges for other dimensions vary with the height and square of the thickness. Charges may be calculated by the formula $c = 3hT^2$ in which c equals the charge in lbs., h the height, and T equals thickness in ft. If the timbers are circular or square, the charge will be same as for trees, taking the diameter of the round and the side of the square one as d in the formula given. (Pl. 56, Fig. 2.)

432.—When high explosives are not on hand, the bridge torpedo used in the war of the rebellion may be used. It consists of a bolt 8 or 9 in. in length, surrounded by a tin cylinder 2 in. in diameter and filled with powder. The ends of the cylinder are closed by iron washers, and a fuse placed in one end. To use this torpedo, a hole 2¼ in. in diameter is bored in the timber and the torpedo inserted having its center at the center of the timber; it is then exploded by means of a fuse. If necessary, two may be placed in holes bored at right angles to each other. (Fig. 4.)

433.—**Palisading** up to 10 in. in thickness takes 4 lbs. of guncotton for each ft. in length to be destroyed.

434.—**For stockades**, when the timbers are close together— 4 lbs. of guncotton per foot in length.

If the timbers are squared, the blocks of cotton may be fastened to a thin board and placed against the foot of the stockade, but if of rough logs, tie the blocks together so that they may adapt themselves to the form of the timbers.

When stockades are double, and separated by a distance of one yard—25 lbs. of guncotton per foot in length or one charge of 80 lbs. may be placed.

Railway iron stockades are breached by 7 lbs. per foot in length.

In all cases, distribute the charge so as to cover the length desired to be breached.

435.—To cut steel rails, use one block of guncotton, weight, 10 ozs., tie the block against the web of the rail with wire or twine, and, if possible, tamp well with earth, and explode. (Fig. 3.)

To blow a piece of some length out of a rail, arrange two charges of 10 oz. as shown above, placing one on each side of the rail at a distance apart of 5 or 6 ft.; these should be exploded by use of a magneto machine so that the action may be simultaneous in both charges; the section will be blown out and turned on its center, making a large opening. (Fig. 5.)

436.—Switch points may be destroyed by lodging a block of cotton between the outer rail and the pivot end of the switch point, being careful to tamp as completely as possible, ("A" Pl. 51, Fig. 11.) In the case of frogs, one 10 oz. block, placed in the angle of the frog, will destroy the point and render it useless. ("A" Pl. 52, Fig, 1.)

437. To destroy a wooden truss bridge, blow away the main brace of the panel nearest to an abutment or pier; it is really necessary to blow only one side down but it is better to be sure and destroy both sides. If the bridge has an arch of wood besides the truss then destroy the arch on each side. If time is available, the bridge may be burned; to do this, collect brush, etc., and, if possible, place it under one end so as to burn it off and cause the span to drop, but if not possible, build the fire in the center of the span; it will burn through and cause rupture in the center.

In all cases attack span over the deepest water.

438.—To cut wrought iron plates the charge of guncotton in pounds is found from the formula $c = 1.5wt^2$ in which c equals charge, w the width of plate to be cut in feet and t the thickness in inches. The charge must be placed entirely across the plate to be cut.

439.—To calculate the charge necessary to rupture an iron bridge beam or girder, calculate the charge for each separate cross section to be cut and add the results; the sum will be the charge required. The charge is most conveniently placed on the side of the beam, reaching entirely across and bound on with wire, the primer being in the center of the charge. If possible, a board should be tied over the whole charge and earth tamped in around it. When time is not available for the placing of the charge as described above, then it may be placed on top of the beam or on the flange, the charge being calculated by the above formula. In demolishing iron girder bridges the charge is applied as described for iron girders. (Figs. 6 and 7.) When the girder bridges are only short ones up to 20 ft. in length, they may be overturned by levers and thrown down from the abutment.

440. When it is necessary to destroy truss bridges the most favorable place for the charge is at the center of the span on the lower chord. When the bridge is of the variety known as a deck bridge, the charge should be placed on the top member, the amount necessary being calculated as shown above. When the lower or tension member is composed of eye bars the charge should be placed between alternate pairs of eye bars as *near to the coupling pin as possible*. (Pl. 57, Fig. 1.) When bridges are supported by iron or wood piers, it is sometimes possible to destroy the piers and thus bring down the entire structure. The charges may be calculated by formulas for iron or wood as the case may be. It is best to destroy the supports on both sides of the pier as this will infallibly bring down the entire structure.

441. **In destroying suspension bridges,** blow down the towers below the saddle, excavate and blow out one of the anchorages or cut the cable with guncotton. The charge for bridge cables is calculated as for cutting iron plates, but the charge must be very carefully placed so as to be in close contact with the cable. If the cables are made of plates, *the plates may be cut with cotton*, as in the case of eye bars, cited above. Where bridges of iron or wood are supported by masonry piers, the piers may be destroyed by charges calculated by the formulas given below.

442.—**To blow down a wall,** the charge should be placed against the bottom as close to the wall as possible, and if possible,

PLATE 57.

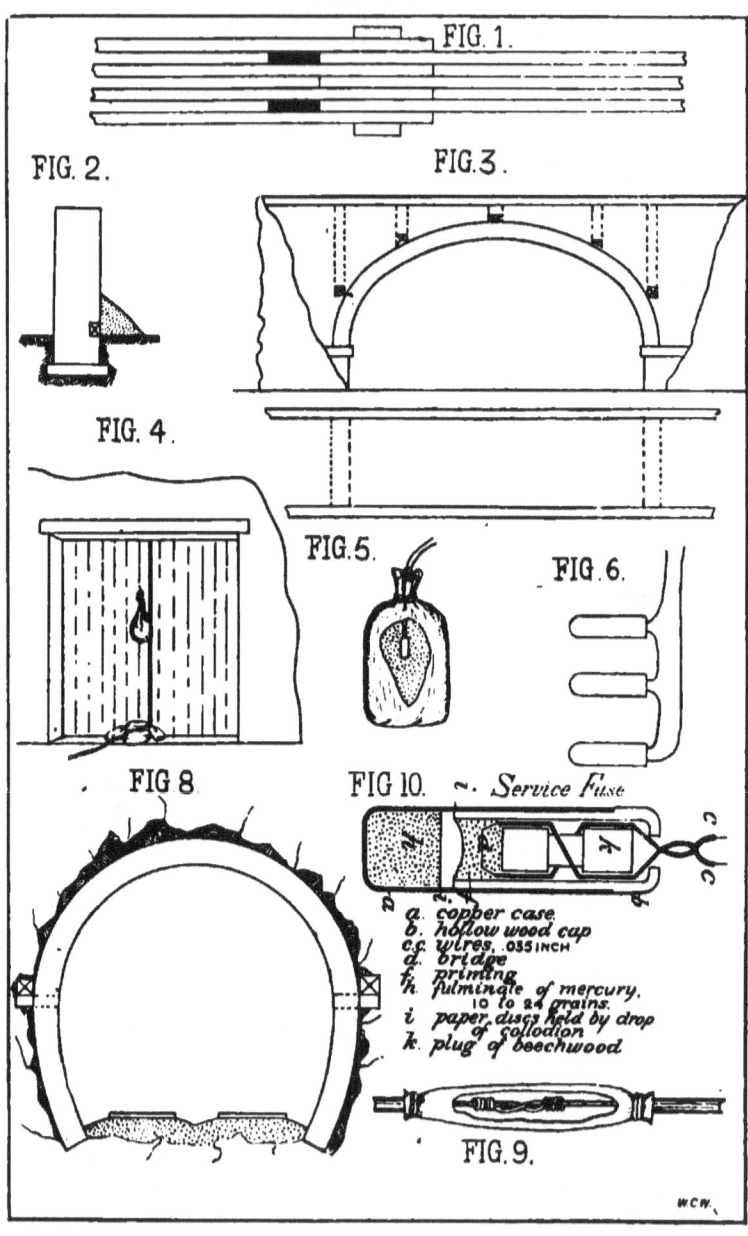

should be tamped. If time allows, a channel should be cut in the wall at the seat of the charge. When untamped, $c = \frac{1}{2}wt^2$ where c equals charge in pounds, w width of wall to be breached, and t thickness, w and t being in feet; and when tamped with earth equal in thickness to the wall, half the above charge may be used. (Fig. 2.)

443.—Bridge piers in masonry may be destroyed by charge of $\frac{2}{3} Wt^2$ ~~per running foot~~. This means excessive expenditure of explosive and would not be used against short, thick piers. In case one has to be destroyed, place small charges in chambers excavated as deep as possible in the masonry, and explode all simultaneously; calculate charges as for walls. Masonry bridges should be attacked at the haunches; if that is not possible, then blow in the crown. When attacking the crown, place explosive in trenches clear across the bridge, and, if possible, on the masonry of the arch itself; better use two trenches, each placed, from the middle of the bridge, a distance equal to ¼ the width to be breached at the crown or arch; calculate charge for each trench clear across the roadway. If haunches are attacked, dig down as far as desired and place charge clear across on back of masonry; charge $\frac{2}{3} Wt^2$ if untamped, and half of this if tamped. (Fig. 3.)

444.—Gates may be blown in with 50 lbs. of guncotton exploded in one charge. The charge is hung against the center of gate by means of a sharpened pick or on a nail. This charge is large, but gates will generally be strengthened in some manner on the inside. (Fig. 4.)

445.—Houses may be blown down or shattered by placing charges in the center of the floor and closing all outlets such as doors, windows, etc. Charge according to the size, 20 lbs. being sufficient for small houses.

446.—Tunnels may be destroyed by placing charges back of the masonry at the spring of the arch. If possible the charge should be placed in a chamber excavated behind the arch and well tamped. Tunnels should be blown in at several places, so as to render it impossible to repair them in a short time. Charges as for bridge arches. (Fig. 8.)

447.—One and one-half pounds guncotton detonated on the chase will disable field or siege guns; the damage is more considerable when the charge is tamped. If not damaged by the first explosion, double the charge and explode in the same place. Heavy fortress guns may be disabled by destroying their carriages by the use of guncotton. Wherever gunpowder is used as an explosive it should be enclosed in stout bags, preferably two, and the outer one well tarred. This is to protect from accidental explosion set up by sparks from the fuse. (Fig. 5.)

448.—The following table gives in a concise form all information necessary for the use of guncotton and gunpowder. The table shows approximately the value of the different high explosives as compared with guncotton.

NOTE:—Charges are in lbs.; B and T are in feet; t is in inches.

B is length of breach to be made; T or t is thickness of object to be demolished.

Gunpowder is assumed to be roughly tamped with sand-bags. Guncotton is untamped. If the guncotton is tamped, the charges may be reduced by about one-half.

Charge of guncotton must be equal in length to the breach which is to be made.

HASTY DEMOLITIONS.

Object attacked	Gunpowder	Guncotton	Remarks
Brick arch......... Brick wall, 2 ft. or less........... Brick wall over 2 ft. thick.........:... Brick piers........	$\frac{3}{8}BT^2$	$\frac{3}{4}BT^2$ 2 lbs. per foot run $\frac{1}{4}BT^2$ $\frac{3}{8}BT^2$	The length of breach, B, should not be less than the height of the wall to be brought down.
Hard wood (e. g., teak, oak, elm), in any form, whether stockade, palisade, single timbers, trees, etc...........	40 to 100 lbs. for stockade	$\frac{3}{8}BT^2$ $\frac{3}{8}T^2$	In a concentrated charge, or for trees not over 12 in. diameter, in a necklace. In auger hole, when the timber is not perfectly round, T = the smaller axis.
Soft wood............	Half the for hard	charges wood	
Breastwork of horizontal balks, or earth, between sleepers, up to 3 ft. 6 in. thick.........	60 to 80 lbs. per 5 ft.	4 lbs. per foot	
Heavy rail stockade..	....	7 lbs. per foot	
Fortress gate	200 lbs.	50 lbs.	
Iron plate	....	$\frac{3}{8} Bt^2$	In this case *only*, t is in inches.
Field or Siege guns..	....	1¼ lbs.	On chase near muzzle.
Heavier guns	....	4 lbs.	In bottom of bore, tamped with water or sand.
First-class iron rail...	....	⅜ lbs.	Touching web of rail and near a chair.
" " steel rail..	....	4 ozs.	Four rails placed round the charge will be cut simultaneously by it.

DEMOLITIONS.

APPROXIMATE RELATIVE STRENGTH OF SOME OF THE HIGH EXPLOSIVES.

Explosive gelatine 128.3
Nitro-glycerine 120.3
Guncotton 100
Dynamite, No. 1, 75% Nitro-glycerine 97.8
Rack-a-Rock 74.2
Dynamite, 50% 72.7

449.—Destruction of Obstacles. Wire entanglements may be destroyed by cutting with wire nippers, or, if they are not at hand, then the ordinary hand axe will do, taking care to cut against the picket.

450. Abatis is very difficult to destroy and cannot be removed while fire can be brought to bear on the spot. Pry up the pickets with levers and attach ropes to the butts of trees and haul away.

451. Palisades and stockades are cut away with axes or saws, the cuts being made near the bottom and ropes being attached to the tops of the timbers; or dig out the earth at the bottom and pull them over.

452.—Small pickets are cut through with the axe, or, if possible, pulled up.

453. Small pits are filled with earth, brush, or covered by planks, fascines, or bales of hay.

454.—Automatic torpedoes are easily destroyed by driving animals up and down the line suspected of containing them.

CHAPTER XXI.—Camping Expedients.

455.—There are a few general principles which should be observed in selecting a camp, whether the troops are to be established in bivouac, in tents or in huts. These principles relate to the health and comfort of the troops, the facilities for communication, the conveniences of wood and water, and the resources of the locality in provisions and forage.

456.—For an intrenched camp the ground must be selected with particular reference to its adaptability for defense and the camp arranged with that object in view, at the same time observing as many of the other requirements as possible.

457.—**Dry and healthy sites, dependent on soil:** *Granite, metamorphic and trap rocks*, usually; *clay slates*, but drinking water is scarce; *limestone* generally, but the water is hard, clear and sparkling though sometimes contaminated; *deep permeable sandstones*, if the air and soil are dry; *deep gravels*, unless lower than surrounding country; *pure sand*, deep and free from organic matter; *well cultivated soils* generally; *gravelly hillocks*, the very best.

458.—**Unhealthy sites, dependent on soil:** *Magnesium limestone; shallow sandstone* underlaid with clay; *clay and alluvial soils* generally; *rice fields; made soils* usually; *newly plowed ground.*

459.—**Healthy sites, independent of soil:** *The best is on a divide or saddle,* unless too much exposed or without water. *The next best is near the top of a slope* and the *southern side is preferable to the northern; banks of running rivers are good,* if not marshy.

460.—**Unhealthy sites, independent of soil:** *Enclosed valleys, ravines,* or *the mouths of long ravines, ill-drained ground, the neighborhood of marshes,* especially if the wind blows from them. If forced to camp near a marsh the windward side should be selected and, if possible, have a hill or a screen of woods or brush between the camp and marsh. Moss generally indicates marshy ground.

461.—Sites affected by surrounding vegetation: Herbage, or closely lying grass, is always healthy, but should be kept cut and all weeds destroyed; heavy brush about a marsh should not be removed. Trees, in cold countries, break the winds; in hot countries, they cool the ground and may protect against malarial currents; so should only be removed with judgment.

462.—In selecting camps, wood, water and grass should be secured if possible, together with good drainage, but marshy ground should not be occupied even for a night.

Old camp grounds should never be occupied if avoidable; instead, go as far as possible to the windward side of them.

The site having been selected, the details of castrametation, or the laying out of camps for the different arms will be found in the authorized Drill Regulations of each.

463.—Water is more immediately necessary to life than food.

Each man requires on the march for drinking and cooking 3 to 4 quarts per day, and an equal amount for washing. In stationary camps 5 gallons per day for all purposes.

Hospitals require several times as much per man per day.

Horses, mules and cattle require from 6 to 10 gallons each per day for drinking. It should be soft and clean, if possible.

Sheep and hogs require from 2 to 4 quarts each, per day.

464.—*It is imperative, on going into camp, that the supply be immediately looked after and a guard placed over it.* If the supply be small, special precautions must be taken and an officer put in charge.

465.—Good drinking water should be bright, colorless, odorless, free from sediment, of pleasant and sparkling taste.

Rain water, collected from a clean surface, after the atmosphere has been well washed, is the purest in nature. *Springs* whose origins are remote from habitations, streams flowing through uninhabited regions, and large lakes, furnish the next best sources of supply.

466.—*If the supply be from a lake, pond, or stream*, separate places for obtaining water for men and animals must be marked out, and care taken that the margin is not trampled into mud and the water made turbid. Where this is likely to occur when animals are watered direct from the source of supply, a hard bottom should be formed for them to stand on, and a barrier formed to prevent them going out too far.

It is better, when convenient, to arrange rows of sunken half barrels, or board troughs raised above the ground, into which the water can be drawn. If the supply be limited, it may be necessary to connect the troughs to prevent waste; if not limited, each should be supplied direct from the source and the overflow drained off. Even when drinking from a running stream, the animals below get foul water. To prevent the ground around the troughs becoming muddy it should be paved and drained along the whole length and for a distance of 10 or 12 ft. back. Where troughs cannot be constructed, trenches lined with puddled clay may be made to answer.

Arrangements should be made so animals may be brought up from one direction and leave in another without confusion or crowding.

467.—*If the supply be from a stream*, the water for drinking and cooking for the troops is drawn highest up; for the animals to drink, next below; and for washing, bathing, etc., lowest down; while all drainage should enter below where any water is taken.

If the stream be small it may be necessary to construct a series of small reservoirs by building small dams across. Animals drink better and more rapidly where water is 5 or 6 in. deep.

468.—If unavoidable, water from small ponds and shallow wells should only be used after being boiled half an hour, then aerated and filtered.

469.—*If the supply be from springs*, each should be enlarged and surrounded by a low puddled wall, to keep out surface drainage. They may be lined with casks or barrels charred inside, or gabions, afterwards working in puddled clay between the earth and linings. The overflow may be received into a succession of casks let into the ground close together.

Surface springs should be sought for in hollows, at the foot of hills, where the earth is moist, the grass unusually green, or the thickest mists arise mornings and evenings.

470.—If water is not immediately available it may be necessary to dig wells. The most expeditious means of doing so is to use *Well Augers*. (Pl. 58, Figs. 1, 2 and 3.)

471.—To dig a well, an auger is attached to a rod suspended from a rope passing over a pulley at the top of a derrick or tripod and thence to a windlass. To the auger rod is secured an arm or arms by which the auger is turned by hand and so screwed down into the earth. About eight turns fills the auger, which is then lifted, emptied and replaced.

472. *The auger for boring in quicksand* (Fig. 3) is shaped similarly to the ordinary wood-boring auger, but with a hollow shank, so that, when lifted, no suction is produced. When the thread becomes loaded the auger is drawn up into an enclosing cylinder, removed from well and emptied.

473.—Driven wells. The driven tube-well consists of a tube about 3 ft. long, perforated with holes, and furnished with a steel point of bulbous form (Fig. 4) and as many other plain iron tubes as may be necessary.

The form of the point serves to clear a passage for the sockets by which the tubes are screwed together.

474.—To drive a well, a tube is screwed to the point (Fig. 5) and on this a clamp is fastened by two bolts at about 3 ft. from the lower extremity of the point. Next, an iron driving weight, or monkey, is slipped on the tube above the clamp. The tube thus furnished is raised and held vertically in the center of a guide in which it is retained by a latch. The whole being now arranged in position, ropes are made fast to the monkey and passed over pulleys on the guide, and driving commenced by two men pulling the ropes and allowing the monkey to fall on the clamp. As soon as the clamp reaches the ground, the monkey is raised and held up, the clamp loosened and raised 1.5 or 2 ft., tightened, and the driving continued as before until the top of the tube comes below the hole in the top of the guide head, when the lengthening bar, (Fig. 7) is dropped into the top of the well-tube. The lengthening bar consists of a length of the well-tubing with a smaller pipe brazed into one end and projecting about 1 ft., which fits into the well-tube. This bar keeps the tube steady and serves as a guide for the monkey to slide on until the top of the well-tube reaches

PLATE 58.

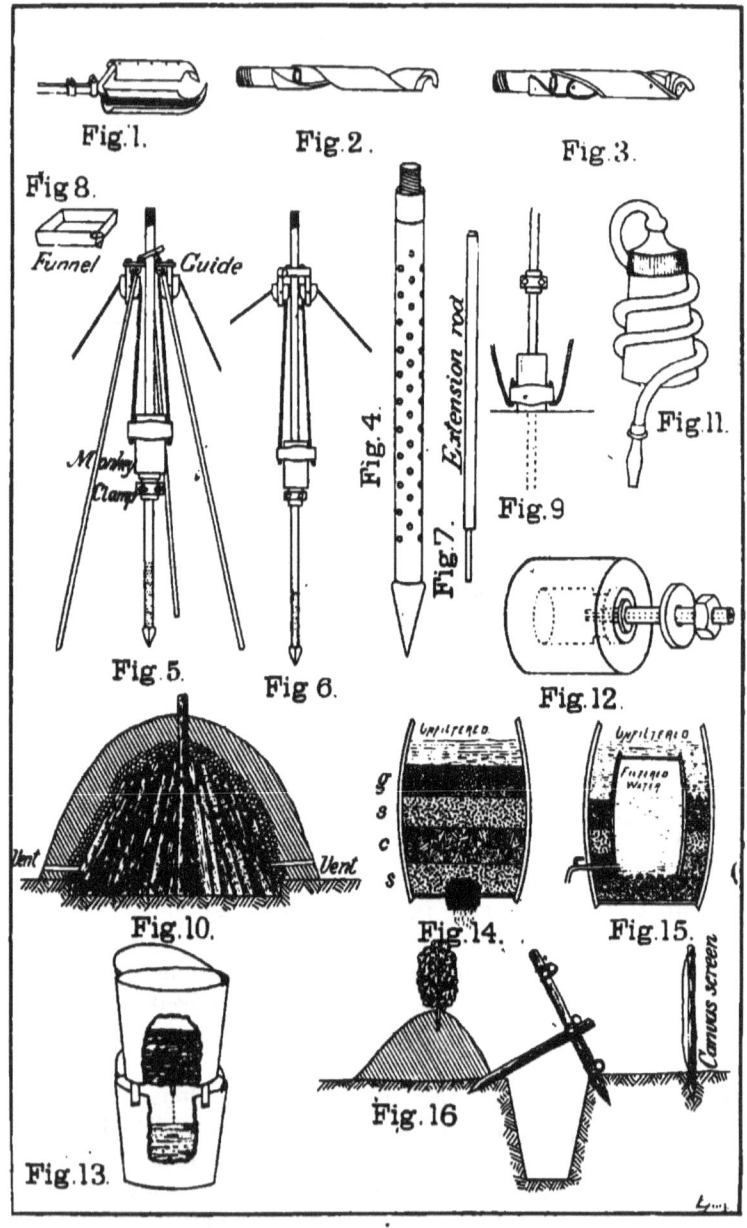

to within a foot of the ground. The lengthening bar is then removed, another tube is screwed on, and the driving continued until water is reached. A hollow iron plumb is frequently lowered into the tube to ascertain when water has been reached or whether earth of any kind has got into it.

Accumulations in the tube, of a loose sandy nature, can be pumped up, by screwing a funnel (Fig. 8) on top of the tube, then lowering into it through the funnel a smaller tube with a pump attached. Water poured into the funnel runs down outside the smaller tube and is pumped up through it bringing the mud and sand. When water is struck, and stands several feet in the tube, the pump is screwed on to the well tube.

The well can also be driven without the use of the tripod supports (Fig. 6) care being taken to keep the tube vertical by means of guy ropes. Such a well can be driven from 10 to 20 ft. per hour. The tubes can be withdrawn without damage by reversing the operations of driving. (Fig. 9.)

475.—The tube-well is not intended for piercing rock, or solid stone formations, but is quite capable of penetrating very hard and compact soils. When striking rock, stone, or deep beds of clay, it is best to pull up the tube and try in another spot, for by going a little distance off water will in many cases be found.

476.—Clarification of water. Water usually contains mineral and organic substances *in solution* and *in suspension*. Substances in solution completely disappear and cannot be entirely filtered out. Substances in suspension do not entirely disappear and may be filtered out.

477.—*Hard water contains one or more substances*, as lime, magnesia, iron and others, *in solution*, which are liable to produce intestinal troubles to persons unaccustomed to them. Cooking vegetables in it is very difficult. Washing with it requires a great deal of soap.

The hardness of water may be partially removed by boiling for half an hour or so, or by adding a small quantity of washing soda, or by adding a couple of ounces of quicklime to 100 gals.

478.—*Substances in suspension* may be largely removed *by precipitation* and *filtration*.

479.—Precipitation is allowing such matter as will to settle through its greater specific gravity, or, by inducing it to do so through some harmless chemical or mechanical action. For which purpose may be used about 6 grains of crystallized alum to the gallon, or tannin in small quantities and letting stand several hours before using; bruised cactus leaves, also tea leaves that have been used act similarly; citric acid, 1 oz. to 16 gal., or borax and alum, ¼ oz. each, or 1 to 2 tablespoonfuls of ground mustard to a barrel improves water.

480. **Filtration** is mechanically arresting and attracting suspended matter, and removing dissolved matter in the water. Filtering materials act only for a short period and should be frequently cleaned.

481.—*Materials which may be used* are sponge, wool, and like articles for straining, but must be constantly removed and cleaned. Clean sand, gravel, and porous stone remove suspended matter, but have little or no effect on dissolved organic matter. *Iron sponge*, a compound of sawdust and iron oxide heated in a furnace, and *Carferal*, a composition of charcoal, iron and clay, are efficient for removing mineral matter. Bone black or animal charcoal, and wood charcoal, when freshly burned, absorb mineral matter for a couple of weeks, but their chief action is on organic matter.

482.—**Charcoal** may be made by digging in the ground a circular pit, some 6 in. deep by 4 or 5 ft. across, then placing a large pole or bundle of brushwood vertically in the center. Around this the wood to be burned is piled, forming a kind of cone. The pile is then covered with brush, and on this a layer of 4 or 5 in. of earth. (Fig. 10.) The center pole is then removed and a fire lighted in its place, receiving air from vents left at the bottom for that purpose. The fire proceeds from the center outwards, and, if burning properly, the smoke is thick and white. If it does not spread to every part new vents must be made. If the smoke becomes thin and a blue flame appears, it is burning too fast, and vents must be stopped up or more earth thrown on. When the smoke ceases to escape, the vents and chimney are closed and the pile allowed to stand for a couple of days until it cools.

From 20% to 25% of charcoal is thus obtained.

483.—**A convenient portable filter** (Fig. 11) is made by taking a small cylinder of compressed carbon and inserting it in a rubber tube in such a manner that the carbon end may be immersed in the water, then applying the mouth to a mouthpiece at the other end of the tube, and drawing the water through.

484.—The Success Filter, (Fig. 12) consists of a cylindrical porous stone 4 in. long by 4 in. in diameter with a hole bored in one end. In this hole is fitted a rubber gasket through which passes an iron tube that is fastened into the bottom of a barrel, jar or bucket. The water filters through the stone into the hole inside and passes out through the tube into a receiving vessel below. (Fig. 13.)

By fastening the iron tube into the bottom of a large empty tomato or peach can, in which the stone is placed on the tube and wedged fast, then fastening a rubber tube 2 or 3 ft. long on the iron tube outside of the can, a syphon filter is obtained. The action is set up by exhausting the air from the stone, after the can and stone are immersed in water, by sucking on the end of the rubber tube until the water is started.

485.—A simple water filter may be made by stuffing a piece of sponge in a hole in the bottom of a cask, flower pot, or other vessel, (Fig. 14) then placing above this a layer of coarse sand, then a layer of pounded charcoal, 3 or 4 in. thick, then another layer of coarse sand, and on this a layer of coarse gravel. The layers should be thick as possible, and tightly compressed, and washed thoroughly clean before being used. The different layers may be prevented from mixing by perforated boards, or otherwise. Another form may be made as shown in Fig. 15.

486.—Casks, or barrels, charred on the inside (and occasionally cleaned, brushed, and recharred) improves water.

487.—Latrines. Arriving on the site of a camp, one of the first duties is to designate the places to attend to the calls of nature, and there dig latrines. Urinals should be placed nearer the camp and of easy access.

The only exception to digging latrines is when the command is very small, is certain to march the next day, and no other troops are to follow.

488.—Latrines and urinals should be so placed as not to be in the course of the prevailing winds to the camp, and must be so situated that they cannot pollute the water, either directly or by soakage.

489.—A small, shallow trench will suffice for a single night, and should invariably be filled in the morning before marching.

For longer periods, a trench 2 or 3 ft. wide at top, from 2 to 10 ft. deep, and 12 to 15 ft. long for every 100 men, should be dug, throwing the earth to the rear, from which a layer of a few inches should be thrown into the trench every day, or oftener if necessary. Lime or charcoal may also be used to deodorize the soil.

It is better to increase the number of trenches than to make any one trench too long.

Shallow latrines should be discarded when filled within a foot of the surface, and completely filled in with earth; deep ones when within 3 or 4 ft. of the surface.

All latrines should be filled in and marked before marching.

490.—In temporary camps, latrines may be provided with seats of a pole and a back, and be screened by bushes, canvas, or other means. (Fig. 16.)

491.—Kitchens. On going into camp, kitchens should be promptly established and in the same relative positions as if the camp were going to last a month or more. A pit should be dug near by for strictly liquid refuse, while solid matter should be placed in a box or barrel for the police party to remove.

492.—*When fuel is plentiful*, a trench of sufficient length and about 1 ft. deep may be dug to contain the fire over which the kettles are hung from supports. (Pl. 59, Fig. 1.)

If fuel is scarce, then dig a trench as above in the direction of the wind but a little narrower than the diameter of the kettles to be used. Place the kettles over the trench and fill in between with stones, clay, etc., forming a kind of flue. The draft may be increased by building a chimney of sods, stones, etc., on the leeward end and enlarging the windward end. (Fig. 2.)

If the camp is to be for a long time and the direction of the wind liable to vary, a number of such trenches may be dug radiating from a common point, over which point a chimney is constructed. Then, whatever the direction of the wind, the trench opening in that direction can be used, the others being closed.

The trenches should have a slight fall from the chimney back for drainage, and a means for the water to escape. If the kettles are small or of various sizes, rests of stones, scraps of iron, etc., may be placed across the trench.

A square hole may be dug for the fire with trenches for draught

PLATE 59.

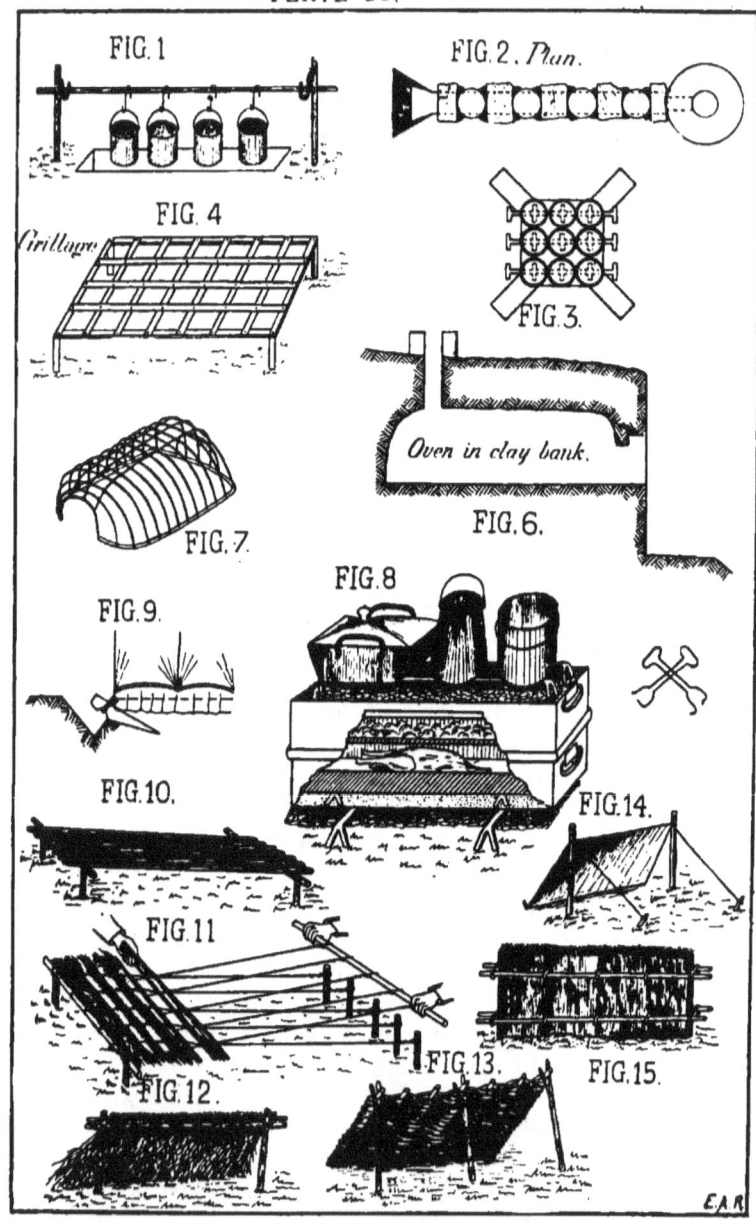

at the corners, the kettles being placed on rests over the fire. (Fig. 3.)

493.—*A grillage* or *kind of grate* about 1 ft. high, made of gas-pipe or bar-iron, is sometimes used to set over the fire, and on this are placed the kettles. (Fig. 4.) These are sometimes made with movable joints so as to be closed for transportation.

494.—If a covered kitchen is desired, either a trench similar to Fig. 2, can be dug, or one above ground can be built with stones and sods and a tent placed over it, or a cover constructed.

495.—To bake bread, when none of the portable ovens of the Commissary Department are carried, improvised ovens must be constructed. The simplest method is to take a barrel with one head out (one with iron hoops best), lay it on its side in a hollow in the ground and then plaster over with wet clay 6 to 8 in. thick, then with a layer of dry earth equally thick, leaving an opening of 3 or 4 in. at the top of the closed end for a flue. The staves are then burned out by a hot fire which also bakes the clay covering, forming an arched oven. To bake, after heating, the front and flues are closed. Or a pit may be dug from 6 to 12 in. deep and 4 by 5 ft. for the hearth, over this form an arch with a hurdle or any other material available (Fig. 7), with a chimney at one end and a door at the other. Then plaster and cover the arch as in the barrel oven and bake the clay covering.

496.—An oven may be excavated in a clay bank (Fig. 6) and used at once.

497.—The Buzzacott Army Field Oven (Fig. 8) which is an article of issue by the Commissary Department, is a complete camp cooking outfit consisting of oven, baking and frying pans, etc. All are securely packed togther and can be conveniently carried in the feed box of an army wagon or on a pack animal. To use it, a bed of live coals is first obtained, then the oven after being heated is placed on rests over a bed of the coals, and a layer of sand sprinkled evenly over the bottom of the oven to prevent burning out. In this is placed the pans of prepared food on suitable rests and the whole covered with a hood. On the hood is scattered a layer of live coals and burning brands. Broiling, frying, coffee making, etc., may be done on top of the oven by using the remaining pans, rests, etc., at the same time that the interior is used for baking.

498.—Drainage. The camp being located, a system of surface drainage should be carefully traced and constructed. As soon as a tent is pitched it should be surrounded by a shallow ditch outside, emptying into a company ditch. The proper method of doing this is to have the inner edge of the ditch come just inside of the skirt wall of the tent to catch the water running down the side of the tent and to drain the interior.

The picket should be inside of the ditch. (Fig. 9.) To bank earth up against the tent soon rots the bottom of the wall.

499.—Beds. The ground being generally too damp to lie upon directly, all should sleep upon some dry material as straw, leaves, or preferably a low platform constructed of small branches and poles, if available. (Fig. 10.)

If required to sleep upon the ground, one will sleep more comfortably if he scrapes out a small hollow for his hips. Straw, hay, etc., for sleeping upon may be made into mats with the Malay Hitch as in Fig 11.

500.—Windbreaks. When troops bivouac, some protection from wind may be obtained by building up to the windward a pile of earth, sods, etc. Where trees are available, by resting a pole on two forked sticks, 4 or 5 ft. high, against which branches, thick end up, are piled at an angle of 45° on the windward side. (Fig. 12.) Hurdles similarly placed, supported and covered (Fig. 13), canvas or blankets secured as in Fig. 14, straw or hay clamped between poles as in Fig. 15, may be used.

By throwing up either a half or whole circle of earth 18 ft. in diameter, from a ditch on the outside, some protection may be obtained. On the bank so formed additional wind-breaks may be placed or a covering extended over it all may be made. (Pl. 60, Figs. 1 and 2.)

501.—In cases of prolonged occupation, if tents are not available, the troops should build shelter of some kind.

Huts may be built of timber, logs, brushwood, adobe, etc., in connection with straw, bark, sods and similar materials.

All huts should, if possible, have their floors raised above the ground to allow free circulation of air underneath. (Fig. 3.) Only in very dry soil and when not to be occupied long is it allowable to sink them, if avoidable. Space between huts in the same row should equal the height of walls, and passage in rear equal the height of ridge. Hut sites should be well pounded.

PLATE 60.

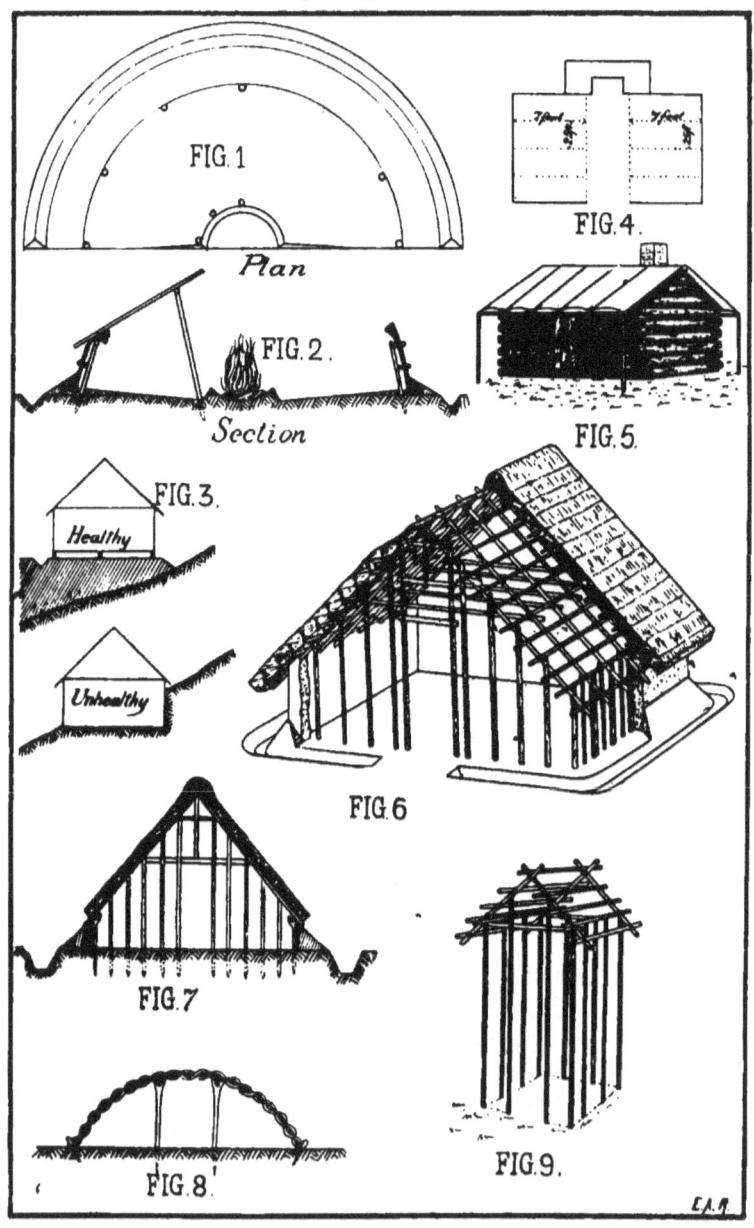

CAMPING EXPEDIENTS. 273

Huts are ordinarily constructed to contain a small number of men, but the sizes and details of construction will depend greatly upon the site and materials available.

A very fair minimum allowance per man of bed space is about 2.5 ft. x 7 ft., with a passage at foot from 2 to 4 ft.

Thus, the plan for 8 men may be taken at 10 ft. x 18 ft., arranged as in Fig. 4. For 12 men, 15 ft. x 18 ft. For 16 men, 20 ft. x 18 ft. For 20 men, 25 ft. x 18 ft.

For calculating the accommodation at the above rates, allow one man per pace of length for a single row of beds and 2 men per pace of length for a double row of beds.

502.—Major Smart, Medical Department, recommends as best a modification of the Army of the Potomac hut, of rectangular plan (Fig. 5), 7 ft. x 13 ft., height to eaves 6 ft., to ridge 10 ft.; door in middle of one long side, chimney opposite door on outside of wall; on each side of doorway a double bunk. This hut to accommodate 4 men.

If logs are used, the ends are trimmed with an axe where they lap at the corners, so they will lie one upon the other throughout their length.

503.—If made of small timber, some style, as in Fig. 6, with thatched roof, might be used.

If, for any reason, it is not desirable to build huts as above, forms may be used as shown in Fig. 7, or hurdles, as in Fig. 8.

504.—Sentry boxes may be made as in Fig. 9, the side covering consisting of watling described in Chap. IX. and the roof thatched.

INDEX.

	Par.
Abatis, consists of, how made...	50
in shallow ditch, of small branches, in front of glacis.	51
how destroyed	450
Advanced Post, rear of.	204
Anchors, number of, scarcity of.	343
substitutes for	344
use of	342
weights of, names of parts of..	341
Angle, equal to a given angle, method of laying out	22
right, method of laying out...	18
re-entrant, salient, shoulder..	82
Approaches, in siege operations constructed by infantry.	154
Area, of rectangle, of trapezoid, of triangle, method of finding.	23
Artillery, in defense of village, where placed	215
in woods	179
projectiles	9
to be placed outside of works	87, 151
Axe, use of...	44
Balks, bay, etc., defined	248
Ballast, for R. R., object of	377
Bank, Gun, definition of, relative advantages of, and embrasures.	70
dimensions of, where placed..	102
space required for	100
Barricade, use and construction of	61
for doors of buildings	195
Batteries, telegraph, how carried	418
Bay, length of, how found	317
Beds, camp.	499

	Par.
Berm, definition of, etc.	65, 99
advantages and disadvantages of	78
Binding, fascines.	117
Bisecting an angle, method of...	20
Blocks, description of, etc., running.	228
Blockhouse, use and construction of	144
in isolated places	145
Boat, buoyancy of how found	316
Box or barrel, to sling	223
ponton, construction of	323
Brackets, telegraph	410
Breaking loads of ropes	218
Breaks in telegraph lines	419
Bridge, anchored to hawser	343
beams, of iron, how destroyed	439
connection of with shore, how made.	350
computing strength of	255
double lock	275
expedients	278
floating, description of	306
flying, description of	309
forming, by successive pontons.	346
forming, by parts	347
forming, by rafts	348
forming, by conversion	349
masonry, how destroyed	443
maximum load for	247
name of, how derived	248, 310
Paine's.	260
pile	269
protection of, from floating objects	352

	Par.
Bridge, continued.	
railroad, repair of	405
requirements of	244
short, how anchored	343
single lock	274
single sling	276
spar railroad	245
suspension	280-6
suspension, how destroyed	441
swing	351
trail	308
treble sling	277
twenty-five feet or less	258-9
twenty-five feet or over	261, 271, 276
Broadside village, how defended	213
Brush huts	503
sentry boxes	504
Brushwood, bundles of, varieties and sizes	115
rate and method of clearing	46
roads	371
Building, defense of, how regarded, first line, how far distant	191
doors of, how barricaded	195
flank defense of	199
how used for defense	189
loopholing of	194
materials used in defense of	200
precautions in defense of	198
removal of	48, 445
requisite of, for defense	190
steps in preparing for defense	193
windows of, how barricaded	197
Buoyancy of casks, how determined	325
Buzzacott oven, description of	497
Cable, charge of explosive to cut	441
swinging, length of	309
Camp beds	499
Camps, drainage of	498
dry and healthy sites for	457, 459
selection of	455, 456, 462
unhealthy sites for	458, 460
windbreaks for	500
Canister, description of	11
Canvas ponton, U. S.	319

	Par.
Canvas raft, description of	303
Capital of field works	82
Caponiers, objections to	85, 146
stockade work used for	188
used in flanking buildings	199
Capstan, description of	212
Casemate, use and general form of	138
how constructed, floor space in	139
Cask, buoyancy of, how determined	325
Casks, closed, piers of, construction of	330
open, piers of, construction of	328
open, safe load of	329
Centrifugal force of train	383
Charcoal, uses of	481
how made	482
Charges, several exploded at same time	429
Chess described	218
Chevaux-de-frise	56
Choker fascine, description and use of	117
Circular village, how defended	214
Clarification of water	476
Clay roads	367
Command of works, definition of	71
Common trench work, how made, uses of	156
Communications, construction of	373, 374
in woods	177
Concentrated load on bridge	253
Conductor, metallic, for telegraph	408
Connecting wires, how done	428
Corduroy roads	370
Counter-scarp, definition, etc.	65, 79
galleries	85
Cover for guns	39
in woods, how obtained	176
Crab	240
Crest, exterior	67
interior	66
military	153
Crib piers, construction of	268
ponton, construction of	322

INDEX. 277

	Par.
Cross, in telegraph wires	419
Cross arms, telegraph	410
Crossing of rivers, selection of, how determined	289
Crossings, railroad	385
Crow's feet	56
Cutting, how defended	171, 172
Dead load on bridges	253
Debris, removal of	48
Defenders of woods, number of	179
Defense, of fences	168
passive, with respect to lines of works	149
Defilade, definition of	88
in plan	89
in section	90, 91
with two planes	92
Depth of fords	290
Derrick, description of	237
in using	240
Destroying railroads, by whom done	399, 400
by whom ordered	401
of telegraph lines	420
Detonator	424
with fuse	426
electrical	427
Digging wells	470, 471, 472
Dimensions of loop holes	164
Disabling railroads	398
Distance between two inaccessible points, method of finding	24
Distributed load	253
Ditch	65, 81, 361
depth of	96
method of digging	101
Doors, how barricaded	195
Double lock bridge	275
Drains, catch and covered	361
Drainage of camps	498
of roads	361
Drinking water	465–9
Driven wells	473–4
Driving piles	270
Dynamite, use of with detonator	429
Earth, excess of at salients	97
in embankments, space occupied by	96

	Par.
Earth, continued.	
loosening of in the front of trenches	38
roads	366
Earthworks, calculation of dimensions of	95
Electrical fuse or detonator	427
Embankment, how defended	170
Embarkation, in ferrying	301
Embrasure	69
used when, details and construction of	103
space required for	100
Engine, locomotive	389
Engineering, Military, Field, definition of	1
Epaulement, gun	39, 70
relative advantage of, and embrasures	70
Equilateral triangle, method of laying out	21
Escape in telegraph lines	419
Escarp	65, 79
Expedients, bridge	278
Exploder, electrical	428
Explosives, kind generally used	421
table of comparative strength of	448
Extending along zig-zag	155
working party	109, 110, 155
Extension on flying sap, method of	158
Eye-bars, how cut	440
Eye splice, to make	222
Faces of works	82
Farms, principles of defense applied to	203
Fascines, size, weight, and making of	117
Fastenings, rail	382
Faults in telegraph lines	419–20
Fences, defense of	168
removal of when	48
Ferry, the rope	307
Ferrying by boat, embarkation, and unloading	301
by raft	302

278   INDEX.

	Par.
Field guns, destruction of	447
range of	13
Field level, description of	26
Filters, portable	483–4
simple	485
Filtering water	480–1
Fire, as regards direction, trajectory	8
double tier of, for walls	167
sector of, discussion of	86
working parties exposed to	111
Flank defense of buildings	199
Flanks of works	82
Floor space in casemates	139
Floating piers, essentials of	314
Flying bridges, raft for	340
telegraph lines	418
sap, description of	157
sap, method of extension along	158
Fords, how made impassable	58
with sandy bottom	291
depth of, requisite of	290
in mountainous country	291
level country	291
how marked, position of, how determined	293
precautions in selecting	295
re-examination of	296
where found	292
Foreground, extent of clearing	43
Form, strap-iron gabion	120
wicker gabion	118
of roads	359
Forming bridge by conversion	349
by parts	347
by rafts	348
by successive pontons	346
Forts, how distinguished	84
Fort Wagner, parallels and approaches to	160
Fortification, classes of	2
compared to other military expedients	7
object of	1
subdivision of field	3
Fougasse, construction, use, and charge for	59
Fraises, construction and use of	55

	Par.
Frogs	384
railroad, how destroyed	436
Fuse	425
how used	426
Gabions, hoop or strap iron, weight and making of	120
method of carrying	157
sheet iron, making of	121
wicker, size, weight, and making of	118
making of without a gabion form	119
Gates, destruction of	444
Gin, description of	239
using	240
lashing, making of	227
Glacis	65, 80
Gorge of works	82
Gradient, limiting of roads	355
Grain, removal of standing	47
Grass, removal of standing	47
Gravel roads	368
Grillage	493
Gauge of railroads	380
Guarding water supply	464
Guncotton	422
how detonated	423
Gun epaulements and pits	39
Gunpowder, how ignited	426
used as an explosive	447
Gutters	361
Hard Water	477
Hasty Demolition, tables showing charges for	448
Head Logs, use of	36
Healthy Camps	457, 459
Hedges, advantages of, how derived, principles of	169
Hedges, removed, when	48
Heights over which fire may be delivered	13
Hitches, knots, etc	219
Holdfasts, description of	243
Hoops, making of, for strap iron gabions	120
Horse, power of on grades	356
Houses, demolition of	445

INDEX.

	Par.
Hurdles, continuous, construction of	123
size, weight and making of	122
Huts, brush	503
how made, etc.	501
Army of Potomac, (Major Smart's)	502
allowance of space in	501
Ice	297
thickness of, how increased	299
thickness of for various loads	298
Infantry Approaches in siege operations	154
Intervals between trenches	34
method of taking by, working parties	109
usual for working parties	111, 112
Insulators, telegraph	412
Insulated wire joints	429
Iron plates, how cut	438
Isolated pits	32
Joints for telegraph wire	417
American twist	417
insulation of	429
rail	381
Junctions, railroad	386
King post truss	272
Kitchens	491-2
covered	494
Knots, hitches, etc	219
Lances, military, telegraph	418
Lashings, gin	227
rack	224
shear	226
transom	225
Latrines	487, 490
Leaks in telegraph line	419
Level, field, description of	26
uses of	27
Lines, cutting of ditch and trench	101
Line, first, falling back past houses	192
parallel to given line, method of constructing	23
Lines, second and third in woods	178

	Par.
Load, distributed, dead, and moving	253
Loading horses in railroad cars	393, 394
Loading wagons on railroad cars	396
Log, cubic contents of	254, 335
buoyancy of	334
Logs, piers of, construction of	338
Long splice, to make	221
Loop holes, dimensions of	164
height of, how influenced	166
how made	36
in buildings	194
Loopholing walls	163
method of	164, 165
Macadam roads	364
Machicoulis gallery, construction of, use of	199
Magazine, general plan of	141, 142
large, small in parapet	140
of gabions and fascines	143
rifle, range, speed, and fire of	14
Magneto exploder	428
Marking fords	293
Materials for bridges	287
for road coverings	365
revetting	114
used in defense of buildings	200
Maximum load for bridge	247
Merlon, definition of, rule as to minimum length of	104
Metal ties	379
Military engineers, duties of	421
Military telegraph lines	418
Mines, land	60
Moving or live loads	253
Objects, floating, protection of bridge from	352
Obstacles, conditions governing use of	49
Office telegraph, treatment of when captured	420
Ovens	495-6
Buzzacott	497
Organizations, or parts of, used as working parties	42
Overhaul tackle	231

	Par.
Paine's bridge	260
Palisades, consist of	54
Palisading, destruction of	433, 451
Palring	118
Pan coupé	102
Paradox, definition of	68
method of determining height of	93
Parallel to a given line, method of constructing	23
Parallels and approaches, Fort Wagner	160
Parapet	63, 67, 81
Perpendicular to a line method of erecting	19
Pickets, forked	119
gabion	118
Piers, bridge, how destroyed	440, 443
Floating essentials of	314
of casks, precautions in using	327
of open boats, precautions in using	315
of open casks, construction of	328
of closed casks, construction of	330, 333
of logs, construction of	338, 339
Pile bridges	269
driving	270
Plank roads	372
Planks, when used for revetments	124
Plows, use of	40
Ponton canvas, U. S.	319
crib, construction of	322
Ponton, box, construction of	323
reserve train, U. S.	320
wagon body, construction of	324
Poles, telegraph, how numbered. telegraph, how guyed, number to mile, preparation of, protection from lightning, raising of, size of, where run	409–411
Portable ramp	393
filters	483–484
truss	279
Position, defensive, definition of, chief requisite of	4
choice of	42

	Par.
Position, continued.	
strength of	153
conditions to be fulfilled	4
of ford, how determined	293
Power of horse on slopes	356
exerted by man	236
of tackle	283–5
Precautions, additional, in defending buildings	198
in fording	295
Profiles, angle, how determined	94
definition and nomenclature of	65
normal, of field works	99
Profiling, method of	94
Pulley, description of	228
Queen post truss	273
Rack, fascine, description and use of	117
lashing to make	224
Raft, canvas, description of	303
for trail bridge	308
of skins	304
Rafts, advantages and disadvantages of	305
for flying bridge	340
swinging, for traffic	351
Rail, fastenings	382
form of	380
joints	381
how cut	435
straightening	407
Railroad, bridge	245
crossings	385
junction	386
wye	387
turntable	388
Railroads, duties of troops in connection with	375
description of	377
destroying and disabling, by whom done	398–400
how disabled and destroyed	402, 403
repair of	404
rolling stock, buildings, etc.	390, 391
Ramp, portable	393
Major Fechet's	397

INDEX. 281

	Par.		Par.
Ramp, continued.		Round timber, strength of	255
semi-permanent	395	Running blocks	228
simple form of	392		
Randing	122	Sag in telegraph wire	414
Redoubts	84	Salient village, how defended	212
Reliefs, 1st, 2nd and 3rd, of working party, cutting lines for tasks of	101	Sand bags, materials, size capacity, and filling	127
		Sap, flying, description of	157
of field work, definition of	72	Saw, teeth of	44
Reserve train, ponton, U. S	320	use of	44
Revetment, definition of	113	Sector of fire, definition of and application to different traces.	86
making and qualities of adobe.	137		
of brushwood	128	Selecting camps	455-6, 462
of fascines	129	Sentry boxes	504
of gabion	130	Sewing, method of, for gabions and hurdles	118
of hurdle and continuous hurdle	131	Shear lashings	226
of pisa	136	Shears, description of	238
of plank	124, 132	method of using	240
of posts	135	Shell	9
of sand bag	134	charges, how exploded	12
of sod	126, 133	shrapnel	10
of timber	125, 132	Short splice, to make	220
Ribbands	55	Shovelers, extra, provided when	101
Road bed	360	Sidings, railroad	384
defense of	173	Single lock bridge	274
materials	365	Single sling bridge	276
Roads, brushwood	371	Site for floating bridge, selection of	311
clay	367		
corduroy	370	plane of, definition	73
desirable conditions in	354	Size of telegraph wire	408
drainage of	361	Slewing	122
earth	366	Slope, banquette	65
form of	359	description of	17
gravel	368	exterior	65, 77
knowledge of	353	interior	65, 75
limiting gradients of	355	superior	65, 76
plank	372	Small pits, how made	57
repair of	369	how destroyed	453
surface of	362	Snatch block	228
width of	358	Sods for revetments, cutting and laying	126
Roadway, weight of, steadiness of	313	Span, superstructure, stringers or balks, side rails, etc	248
width of on bridges	248, 280-6		
Rope, breaking loads of, weight of	218	Spans, 25 ft. or less	258, 259
		25 ft. or over	261, 271, 276
composition, size of, etc	216	Spars, arrangement of	257
Rope, parts of	219, 230	Splice, long	221
rule for strength of	217	eye	222

	Par.
Splinter-proof for trenches....	35
Stockade, advantages of, definition of.................	180
how destroyed 434,	451
kind of timber preferable for.	186
loopholes in..................	184
loopholes, when cut in.......	185
of vertical timbers........	182
of same, square and round timbers	183
of horizontal timber..........	187
Stockade, of R. R. iron, destroyed how	434
when employed..............	181
work used for tambours and caponiers................	188
Stones in trenches, where placed	34
Straightening rails..........	407
Streams, unfordable, how passed	294
velocity of, how determined..	296
width of, how determined....	312
Strength of materials........	249
of rope..................217,	218
Sub-drains	361
Suspension bridges..........280–6	
Switch, split and stub..........	384
Table of breaking loads.........	218
of constants "C"..........	253
of weights of materials.......	256
showing amounts of revetting materials for 100 linear feet of revetment........	137
Tackle, description of..........	229
formula for power of..........	235
power of..................233–4	
to prevent twisting..........	232
to round in, to overhaul......	231
Tambour.......................	147
stockade work used for.......	188
used in flanking buildings ...	199
Tasks.............(Pl. 14, 16) 111,	112
laying out of.....	101
in constructing parallels and approaches............PI. 22	
responsibility for completion of.....................	106
Telegraph messages...........	408
lines, how destroyed.........	420

	Par.
Telephone, outpost cart for.....	418
Telford roads................	363
Terreplein	74
Thickness of materials proof against small arms..........	15
Ties, metal...................	379
wood	378
Timber Bridge, how destroyed	431
felled, removal of..........	45
kinds preferable for stockade	186
round for revetments........	125
standing, removal of .... .. .43,	44
Tools, carrying of by working parties......................	108
cutting, intrenching used in the field	41
taking of, by working parties..	107
used in felling timber........	44
Torpedo, U. S. bridge..........	432
Torpedoes, automatic..........	454
Trace, definition of.............	64
selection of..................	89
Transom, strength of.....	250
Traverse, definition of..........	69
method of determining height of.............	93
Tread banquette..............	65
Treble sling bridge............	277
Tree insulator and tie..........	415
Trees, cutting of.............~.44,	45
how to fell with explosives....	430
Trench..........................	65
common, uses of, how made..	156
drainage of...................	98
method of digging..........,	101
Trenches, advantages and disadvantages of...................	38
disguising location of........	38
intervals in line of...........	34
kneeling or sitting............	30
location of..........	33
loosening earth in front of....	38
lying........................	29
standing	31
Troops, weight of on bridge.....	252
Trestles, capped	264
tie-block	263
two-legged..................	265
three-legged	266

INDEX. 283

	Par.
Trestles, continued.	
four-legged	267
six-legged	262
Truss, king-post	272
queen-post	273
portable	279
Tunnels, how destroyed	446
repair of	406
Turntables, railroad	388
Unloading in ferrying	301
horses from R. R. cars	393, 394
wagons from cars	396
Urinals	488, 489
Velocity of streams, how determined	296
Village, advantages and disadvantages of for defense	206
artillery, where placed in defense of	215
broadside, how defended	213
circular, how defended	214
cover for supports and reserves in	215
in defense of, precautions necessary	205
defense of depends on	208
arrangements for defense of	205
garrison of, how determined	211
objects in holding	207
salient, how defended	212
value of, for defense	205
arrangement for defense of, how made	210
Wagon body ponton, construction of	324
Wagons, prepared for crossing on ice	299
Walls, for double tiers of fire	167
discussion of as military expedients	161
loopholing	163–5
preparation of for defense	162
removal of	48
how destroyed	442, 451
Water, clarification of	476
drinking	465–9
filtering	480–1

	Par.
Water, continued.	
guarding	464
necessity of and amount required	463–4
Water tables on roads	361
Watling	118
Weeds, removal of	47
Weight of materials	256
of rope	218
of troops on bridge	252
Wells, digging of	470–2
driven	473–4
Width of roads	358
Winch	240
Windlass, description of	241
Windbreaks for camps	500
Windows of buildings, how barricaded	197
Wire, connection of	428
entanglements, high, low, how made	52, 53
entanglements, how destroyed	449
how stretched, hanging of, secured to	414
how strung across roads	411
how strung across streams	415
telegraph	408
tie	413
Withes, making and use of	117
Woods, artillery in	179
communications in	177
cover in	176
lying beyond position	175
number of defenders of	179
preparation of edge of	174
2nd and 3rd lines in	178
Works, constructed by troops to occupy	5
double line of	150
fixed types of necessary	6
groups of, intermediate, when used	149
line of	148
advantages and principles of	149
Works, field, conditions to be fulfilled	62
calculation of cross section of	95
classification as to trace	83
defilade of	88, 89, 90, 91, 92

INDEX.

Works, continued.
  details of construction of.99, 100, 101
  length of crest for assumed garrison.................... 100
  open, closed, and half closed, definitions, advantages and disadvantages............ 83
  continuous line of, cremaillere, blunted redan, redan with

Works, continued
  curtains, terraille, terraille and redan, trace of....... 152
Working parties, extension of.109, 110
  organization of............... 105
  when under fire.............. 111
Wye. R. R....................... 387
Zig-zag, direction of, length of. 159
  extending along............. 155

www.ingramcontent.com/pod-product-compliance
Lightning Source LLC
Chambersburg PA
CBHW032104220426
43664CB00008B/1128